UNIVERSITIES IN THE
AGE OF CORPORATE SCIENCE

UNIVERSITIES

in the

AGE *of* CORPORATE SCIENCE

The UC Berkeley–Novartis
Controversy

ALAN P. RUDY, DAWN COPPIN,
JASON KONEFAL, BRADLEY T. SHAW,
TOBY TEN EYCK, CRAIG HARRIS,
AND LAWRENCE BUSCH

TEMPLE UNIVERSITY PRESS
Philadelphia

Temple University Press
1601 North Broad Street
Philadelphia PA 19122
www.temple.edu/tempress

The paper used in this publication meets the requirements
of the American National Standard for Information Sciences—
Permanence of Paper for Printed Library Materials, ANSI Z39.48-1992

Library of Congress Cataloging-in-Publication Data

Universities in the age of corporate science :
the UC Berkeley–Novartis controversy / Alan P. Rudy . . . [et al.].
        p.   cm.
Includes bibliographical references and index.
ISBN 10: 1-59213-533-1 (hardcover : alk. paper)
ISBN 13: 978-1-59213-533-2 (hardcover : alk. paper)
1. Business and education—United States.   2. Education, Higher—
Economic aspects—United States.   3. University of California, Berkeley.
4. Novartis Agricultural Discovery Institute, Inc.   I. Rudy, Alan P.
        LC1085.2.U56 2007
        378.1'03—dc22   2006016232

2 4 6 8 9 7 5 3 1

TODAY, the solitary inventor, tinkering in his shop, has been over shadowed by task forces of scientists in laboratories and testing fields. In the same fashion, the free university, historically the fountainhead of free ideas and scientific discovery, has experienced a revolution in the conduct of research. Partly because of the huge costs involved, a government contract becomes virtually a substitute for intellectual curiosity. For every old blackboard there are now hundreds of new electronic computers.

The prospect of domination of the nation's scholars by Federal employment, project allocations, and the power of money is ever present and is gravely to be regarded.

Yet, in holding scientific research and discovery in respect, as we should, we must also be alert to the equal and opposite danger that public policy could itself become the captive of a scientific-technological elite.

It is the task of statesmanship to mold, to balance, and to integrate these and other forces, new and old, within the principles of our democratic system—ever aiming toward the supreme goals of our free society.

Another factor in maintaining balance involves the element of time. As we peer into society's future, we—you and I, and our government— must avoid the impulse to live only for today, plundering, for our own ease and convenience, the precious resources of tomorrow. We cannot mortgage the material assets of our grandchildren without risking the loss also of their political and spiritual heritage. We want democracy to survive for all generations to come, not to become the insolvent phantom of tomorrow.

—President Dwight D. Eisenhower, "Farewell Address" (1961)

CONTENTS

Preface · *ix*

Acknowledgments · *xi*

Glossary of Acronyms · *xiii*

ONE · Theoretical Framework · *1*

TWO · The Changing World of Universities · *18*

THREE · Land Grant Universities, Agricultural Science,
and UC Berkeley · *35*

FOUR · A Chronology of Events · *46*

FIVE · Points of Contention · *71*

SIX · Overview and Analysis of the Agreement · *83*

SEVEN · The Agreement and the Public Stage · *94*

EIGHT · The Scientific Enterprise · *109*

NINE · Intellectual Property Rights · *137*

TEN · The Impact and Significance of UCB-N on
UCB and CNR · *150*

ELEVEN · Rethinking the Role of Public and
Land Grant Universities · *163*

TWELVE · Constructing the Future: Re-visioning Universities · *179*

Notes · *209*

References · *215*

Index · *233*

PREFACE

HOW PEOPLE FEEL about the promotion and proliferation of university-industry relations at public universities is contingent on many things. Such relations may be seen as the best alternative for pursuing the funds lost when state legislatures reduce appropriations for higher education, whether in absolute or relative terms. They can be understood to suborn the independence and mission of public higher education. They might be viewed as a necessary route to prepare the public and private sectors for regional economic growth in the face of an increasingly global, information-based economy. They are sometimes approached as administratively led patterns of university corporatization that undermine historical patterns of shared governance. They may serve to generate networks and revenue streams for future forms of basic and applied research. They may simply present an opportunity that university administrators and faculty embrace without thinking very much about the long-term consequences. They can be something opposition to which may damage one's career prospects. They may be something whose time has come, but they may also be in decline. They may be all of these things.

When faculty and administrators in the College of Natural Resources at the University of California, Berkeley requested proposals for funding department-wide research in the Department of Plant and Microbial Biology, and then negotiated a deal with the newly formed Novartis Agricultural Discovery Institute (NADI) in 1998, the university exploded in controversy. All of the positions laid out above, and more, entered into the smoldering conversations, emotional debates, and screaming matches that ensued. This explosion had everything to do with the institutional, political, and technoscientific history of UC Berkeley itself, intensified by the Bay Area's history as the heart of the "Left Coast" and California's deeply contested patterns of agricultural development. Coverage of the controversy, while most intense in the Bay Area, was reported in newspapers, professional journals, and national magazines across the country, intensifying the already fraught politics of biotechnology, higher education, and the science wars.

The study presented here resulted from a two-year effort by the Academic Senate at Berkeley, including many faculty opposed to or with reservations about the agreement, to have the university finance an examination by independent researchers of the agreement's history, negotiation, execution, and consequences. Significantly, our work and the conclusions we have reached have proved every bit as contingent as the perspectives people took on the university-industry agreement. One central event here is that the agricultural biotechnology bubble burst—for reasons having effectively nothing to do with the agreement—very soon after the agreement with Novartis was signed. Not only did this eventually lead to the restructuring of NADI into the Torrey Mesa Research Institute (itself subsequently disbanded), it led to Novartis's spinning off its agricultural division to form Syngenta. Another key moment was the publication of a controversial article in *Nature* by an activist faculty member and one of his graduate students about their discovery of genetically modified materials in the DNA of indigenous Mexican landraces of maize. The apparently organized attacks on the article and on its untenured author, Ignacio Chapela, by proponents of agricultural biotechnology also generated a great deal of controversy. Professor Chapela became embroiled in an extraordinary series of events and an extended institutional battle that finally ended when he was granted tenure.

In a world in which the materials of and discourses about the social and the natural, the scientific and the political, the democratic and bureaucratic, the public and the private, and the global and the local are increasingly hard to separate, things like university-industry relations make more and more sense while simultaneously becoming more and more controversial. Our account seeks to present events and arguments in a form sufficiently distilled that they make sense, while at the same time evaluating those events using perspectives oriented to the increasingly hybrid and contested character of contemporary social and political life.

ACKNOWLEDGMENTS

A STUDY OF THIS SORT is always the result of the support and help of a wide variety of people. We wish to take the space here to thank those who enabled this report to be much more than it otherwise might have been. Dr. Anne MacLachlan of the Center for Studies in Higher Education was our designated contact at the University of California, Berkeley, and she more than lived up to her role. Anne provided us with help in locating persons and documents, in identifying paths that might otherwise have been missed, and with amazing insights into the workings of the university. Dr. Jean Lave's willingness to provide space in which to house our research associate, Dr. Dawn Coppin, is greatly appreciated, as is her tenacity in ensuring that this project went ahead.

We also had the good fortune of working with a liaison committee at the University of California, Berkeley, who sought to smooth any bumps we might encounter in conducting this study. In this capacity our thanks go to Dr. Bob Spear, chair, Dr. Jim Evans, Dr. Jean Lave, Dr. Steve Lindow, Dr. Anne MacLachlan, and Dr. Karen de Valois.

It almost goes without saying that this study would not have been possible without the willingness of so many people at Berkeley—faculty, students, staff, administrators—as well as people off campus to take the time to talk with us about their involvement with, and thoughts about, the University of California, Berkeley–Novartis agreement itself. We are deeply grateful for their participation and hope we have not misinterpreted their comments, although the conclusions we draw may not be entirely to their liking.

Dr. Daniel Kleinman and Dr. Jim Fairweather provided helpful detailed comments on various parts of the report upon which this manuscript was built. We thank them for taking the time to review a rather lengthy document. We would also like to thank Dr. Ann Austin for her assistance in the development of this project.

Finally, we are grateful to the University of California, Berkeley, for the funds that made this study possible. Without that generous support, the study would never have been conducted.

Of course, the findings, interpretations, and conclusions drawn here remain the responsibility of the authors.

GLOSSARY OF ACRONYMS

C4	Committee of Four (Wilhelm Gruissem, Bob Buchanan, Peggy Lemaux, Gordon Rausser)
CNR	College of Natural Resources
COI	Conflict of interest
COR	Committee on Research, UC Academic Senate
CRADA	Cooperative Research and Development Agreement
CSHE	Center for Studies in Higher Education
CT	Conventions theory
DANR	Division of Agriculture and Natural Resources
DIVCO	Divisional Council of the Academic Senate
ESPM	Department of Environment Science, Policy, and Management
EVCP	Executive vice chancellor and provost
ExCom	Executive Committee of CNR
FTE	Full-time equivalent
IP	Intellectual property
IPR	Intellectual property rights
LECG	Law and Economics Consulting Group
LGU	Land grant university
MCB	Department of Molecular and Cell Biology

MIT   Massachusetts Institute of Technology

MSU   Michigan State University

NABRI   Novartis Agribusiness Biotechnology Research Institute, Inc.

NADI   Novartis Agricultural Discovery Institute, Inc.

NIH   National Institutes of Health

NSF   National Science Foundation

NST   Department of Nutritional Sciences and Toxicology

OTA   Office of Technology Assessment

OTL   Office of Technology Licensing

PB   Department of Plant Biology

PGEC   Plant Gene Expression Center

PI   Principal investigator

PMB   Department of Plant and Microbial Biology

PPA   Plant Patent Act

PVPA   Plant Variety Protection Act

R&D   Research and development

SAES   State Agricultural Experiment Station(s)

SPO   Sponsored Projects Office

SRR   Students for Responsible Research

TMRI   Torrey Mesa Research Institute

UC   University of California

UCB   University of California, Berkeley

UCB-N   University of California, Berkeley–Novartis agreement

UCOP   University of California, Office of the President

UIR   University-industry relations

USDA   United States Department of Agriculture

WARF   Wisconsin Alumni Research Foundation

UNIVERSITIES IN THE
AGE OF CORPORATE SCIENCE

ONE

Theoretical Framework

MOST PHILOSOPHIES OF EDUCATION appeal to philosophy in general, and to ethics in particular, to argue for (or against) a particular perspective. Of particular importance is the argument that a given ethical perspective is valid across all institutions, places, and times. This unitary view is commonly found in a variety of philosophies, including utilitarian, Aristotelian, and Kantian perspectives.

However, as we interviewed faculty, staff, administrators, and various stakeholders involved in the dispute over the University of California, Berkeley–Novartis agreement, and in our review of a multitude of documents generated by it, we encountered many apologists for the agreement. Not surprisingly, we also found a wide range of critics. Both sides tried to justify their positions. We soon realized that no unitary position would suffice for our inquiry. In contrast, conventions theory, perhaps best described as situated at the junction of philosophy and sociology, appeared to serve as an excellent starting point for our analysis.

Conventions theory (CT) originated in the mid-1980s in the work of Luc Boltanski and Laurent Thévenot, two researchers at the École des Hautes Études en Sciences Sociales in Paris (Boltanski and Thévenot 1991, 1999; Thévenot 2001). It developed in part in response to regulation theory (Aglietta 2000 [1979]; Boyer and Saillard 2002), a structuralist approach to understanding the role of the state in shaping economies, and was also instrumental in shaping actor-network theory (Latour 1987, 1993). Proponents of CT reject both structural and normative approaches in favor of an interpretive approach. Put differently, CT posits *neither* a fixed world in which structural forces (e.g., the state, corporations, classes) more or less inexorably interact in certain ways, *nor* a world in which there are more or less fixed norms to which persons are socialized to adhere under penalty of sanction. Instead, proponents of CT argue that the organization of society is best understood by focusing on the conventions—shared sets of practices—that simultaneously construct, shape, and reshape social relations. To date, CT has been applied mostly in the domain of economics, as an alternative to the methodological individualism of neoclassical models and as a supplement to institutional approaches (for an overview, see Biggart and Beamish 2003). We believe that CT can be applied equally well to the issues facing higher education addressed in this volume.

That said, it is important to note that we attempt here simultaneously to do both more and less than Boltanski and Thévenot have done. Central to the methodology of their project is the construction of an empirically grounded theory of justifications—an explanation for why actors use the justifications they use. Moreover, their empirical data are largely taken from how-to manuals—books and pamphlets that claim to provide guides to achievement in various domains. While we will appeal to some institutional documents that take normative positions that justify specific kinds of policies in higher education, our project here makes no claims to extending conventions theory. At the same time, however, it brings to bear a much wider range of empirical materials with which to examine the justifications and other claims made by various persons and organizations. With this in mind, we begin by examining some of the key insights and claims of conventions theory.

First, conventions theorists, much like interactionists (Clarke 1997; Fujimura 1988; Star 1991; Strauss 1978), note the existence of multiple worlds within any given social setting. These social worlds have no existence apart from the shared practices that order and justify them. Convention theorists argue that such shared practices can themselves be a subject of study since each "world" has its own "orders of worth." These are used in everyday interaction to rank, and are the justifications for ranking of, persons and things. Thus, in the domestic world, a great mother would be someone who cared greatly for her family. She would be ranked higher than would a mother whose care for her family had lapsed. Furthermore, each world contains its own set of more or less accepted "common higher principles." These principles are often appealed to in efforts to settle disputes. Rejecting an idealist and modernist view that insists that all members of an institution follow a common set of abstract principles, proponents of CT insist that the determination of institutional principles be grounded in the analysis of texts, discourse, and other practices common in a given institution—materials that may appeal to a number of different worlds of justification to order action.

Based on their study of iconic texts (i.e., how-to manuals), Boltanski and Thévenot argue that there are six empirically identifiable worlds, although they leave the door open for others. Doubtless, different formulations might identify different numbers of worlds, but the six identified in italics below appear adequate for our analysis. Moreover, the specific ways in which these justifications are applied surely vary from one cultural setting to another over both time and space.

Thus, for example, within the *domestic world* one makes appeals to affective relations, including the importance of supporting one's family members over and above others, and acts accordingly. One rejects economic exchanges among family members, insisting instead upon gift giving. And one takes full account of the special role that parents have with respect to children by treating them differently from others outside the family.

By contrast, in the *industrial world* one appeals to values such as efficiency, precision, and know-how. The most worthy objects and relations in the industrial world are well designed, functional, and conserving of resources invested or transformed. This, in turn, might be compared to the *merchant world,* in which one appeals to how much a person or thing is worth in monetary terms. The salesperson who lands the most sales, the product that sells far better than others—these are the things that are revered.

In the *civic world* one appeals to those practices, persons, and things that serve the public good—often at the expense of family or business. In the *world of inspiration* one appeals to the transcendent, the mysterious, the invisible, as well as to humility. And in the *world of opinion* one appeals to celebrity, publicity, fame.

Within each of these worlds one usually attempts to resolve disputes by appealing to these principles or orders of worth, and by imposing sanctions against those who fail to strive for at least a modicum of worth within that world. Thus the father who shirks the responsibility of caring for his children is perhaps told that he is a poor father. The salesperson who fails to make the sales quota is perhaps sent for further training in salesmanship. The industrialist whose products fail to meet safety standards is fined for the shoddy quality of his or her goods. The celebrity who hides from the public soon is forgotten by the world of public opinion.

Significantly, CT acknowledges the messiness of the real world. While these multiple worlds of justification are made real by their use, their boundaries remain poorly defined, and the worlds themselves frequently collide. While philosophers of various persuasions have traditionally preferred to construct elegant, orderly, and abstract notions of the good, the true, and the beautiful, the world in which we find ourselves tends toward disorder. As we shall elaborate below, CT suggests the need to examine these points of contention carefully, and to do the same with the tests and trials through which worlds, institutions, and practices are made and transformed. Conflicts on the campuses of our institutions of higher education, such as that over the agreement with Novartis that is the subject of this book, are consequences of the emergence of Kerr's "multiversity" in the 1960s, as well as of Delanty's "university in knowledge society" today. Different locations within universities transitioning through these conditions—and different discourses about universities—often appeal to the same ideals but define them by selectively drawing from different worlds of justification. We certainly found this to be the case in our study of the University of California, Berkeley–Novartis agreement.

HIGHER EDUCATION:
CREATIVITY, AUTONOMY, AND DIVERSITY

While a broad definition of higher education might include all forms of post-secondary education as well as related research and service, for the purposes of this volume we limit our discussion to institutions that award bachelor's, master's, and doctoral degrees, and particularly to those universities that engage in teaching, research, and public service activities. Although other kinds of post-secondary institutions play an important role in American society, there is widespread agreement that universities are the leaders in higher education.

As with other worlds of human endeavor, higher education may be said to have its own orders of worth, its own common higher principles. There are arguably three central principles and associated practices that must stand at the center of the world of higher education. We distilled these three principles empirically in somewhat the same way that Boltanski and Thévenot distilled the principles they describe in their works—through a careful review of canonical texts. But we supplemented our textual analysis with numerous interviews, both on and off campus, as well as with a broader reading of the immense literature on higher education. We call these principles *creativity, autonomy,* and *diversity.* Although occasionally obstructed or challenged, these principles are central to the ethical framework of the university. Let us examine each briefly in turn.

Creativity

A central principle—perhaps *the* central principle—of universities is creativity. Creativity is often thought of as the ability to bring something new into existence, to use one's imagination to produce something new. Yet, much like Justice Potter Stewart's famous line about obscenity—"I know it when I see it"—creativity remains to some degree ineffable. Given the scope of modern universities, creativity includes activities ranging from the solution of complex (and hitherto unsolved) mathematical or scientific problems, to the production of written texts that shed light on issues previously overlooked, to the imaginative production or performance of music or drama. Somewhat paradoxically, creativity requires considerable knowledge of existing traditions, such that the creative act may be distinguished by those within a given tradition as truly creative.

Universities are successful as organizations only to the extent that they foster and cherish creativity among their faculty, students, and staff. In the University of California, Berkeley Strategic Plan, it is noted that "Berkeley must provide a research environment that optimizes creativity" (UCB Strategic Planning Committee 2002). Creativity can be nurtured by encouraging freedom of inquiry, but it cannot be created by bureaucratic means. Little evidence suggests that reward systems for scholarly inquiry actually promote an increase in the *quality* of scholarship (although such systems can certainly be used to reward those who succeed).

Part of the problem is that creativity is, by definition, recognized ex post facto (Brannigan 1981). While some scholarly achievements are recognized nearly instantly by others in a given field (consider the case of Watson and Crick's discovery of the structure of DNA), in some instances years or even decades may go by before such contributions are recognized. Examples include Barbara McClintock's "jumping genes" and Alfred Wegener's arguments for what eventually became known as plate tectonics. In the social sciences, humanities, and arts, it is not uncommon for work to be recognized only after its author has passed from the scene.

Indeed, some would argue that reward systems currently in place in large research universities put too much emphasis on quantifiable outputs that diminish the creativity of faculty: numbers of publications in scholarly journals, monetary value of research grant funding, and number of patents received. Some departments in some universities go so far as to stipulate in which scholarly journals one should publish. There is little evidence to suggest that any of this stimulates creativity and quite a few suspicions that it diminishes it.

Indeed, we would argue that creativity can often be thwarted by such bureaucratic means. Incessant formal audits, lengthy and ever more interventionist reviews, and too great an emphasis on receipt of extramural grants, can and do block creativity because they fail to consider that the pursuit of creativity requires risk taking and failure. The AAUP used this as a justification for its reluctance to embrace the currently popular idea of post-tenure review (American Association of University Professors 2005a). From the perspective of conventions theory, the AAUP used a justification from the academic world to argue against the intrusion of a practice that might well be justified in the industrial world.

Creative scholars bring something new into the world. Doing so almost always requires tinkering, a process of trial and error that is concealed in the final paper, book, or artifact (Knorr-Cetina 1981). Formal audits are far too closely linked to a preconceived language and logic of short-term productivity to foster creativity (Power 1997). They ignore the dead ends, the mistakes, the "wasted time" not engaged in "productive" work that is necessary to achieve quality in academic endeavors (Giri 2000). Moreover, such audits raise fundamental questions as to who has the power and right to judge and who will judge the legitimacy of the judges (Bourdieu 1997). We return to the issue of auditing academia in Chapter 2.

Creativity can also be thwarted by tying university research and teaching too closely to the needs of various stakeholder groups. President Eisenhower warned of the undue influence of government on universities, while recent concerns have been focused on the corporate world. A recent AAUP report approvingly quoted the president of Columbia University, George Rupp, as follows: "The danger exists that universities will be so assimilated into society that we will no longer be the kind of collectors of talent that allow creativity to blossom. We must guard against being harnessed directly to social purposes in any way that undermines the fundamental character of the university" (American Association of University Professors 2005b). The authors of that report specifically cited the agreement discussed in this volume as potentially stifling to faculty creativity.

Autonomy

In order to achieve greatness in the world of knowledge, a substantial degree of autonomy is also necessary. Of course, this autonomy is never absolute, nor should it be. Moreover, the kind of autonomy needed varies considerably across disciplines and is, ironically, always dependent on access to various resources. A philosopher or historian may need the autonomy to consult whatever sources she feels are appropriate to the subject at hand. In contrast, an engineer may need certain critical tools and a laboratory in order to express autonomy. But, however defined, without autonomy universities soon lose their raison d'être. They become bureaucratic entities that perform their tasks in a rote manner.

The principle of autonomy is recognized in the "1940 Statement of Principles on Academic Freedom and Tenure" of the American Association of University Professors, which notes:

> Institutions of higher education are conducted for the common good and not to further the interest of either the individual teacher or the institution as a whole. The common good depends upon the *free search for truth and its free exposition.*
>
> Academic freedom is essential to these purposes and applies to both teaching and research. Freedom in research is fundamental to the advancement of truth. Academic freedom in its teaching aspect is fundamental for the protection of the rights of the teacher in teaching and of the student to freedom in learning. It carries with it duties correlative with rights. (American Association of University Professors 2005a)

The American Federation of Teachers makes a similar argument:

> Faculty and professional staff must be able to exercise independent academic judgment in the conduct of their teaching and research. Academic freedom is important because society needs "safe havens," places where students and scholars can challenge the conventional wisdom of any field—art, science, politics or whatever. This is not a threat to society; it strengthens society. It puts ideas to the test and teaches students to think and defend their ideas. (American Federation of Teachers 2005)

Recognizing that autonomy at the individual level requires autonomy at the organizational level, the Academic Senate at UCB recently noted that institutional autonomy was central to the long-term success of the university (University of California, Berkeley 2003a).

One specific issue pertaining to the principle of autonomy concerns basic research. Often, basic research is seen as somehow removed from society, without direct social consequences, and thereby as more autonomous than applied research, or research funded for a specific purpose. But, as Harding (1991, 38) asks, "Why should society, in the face of competing social needs, provide massive resources for an enterprise that claims itself to have no social consequences? There is a vast irrationality in this kind of argument for the purity of science." Note that Harding does not argue against basic research as such but against the idea that it has no social consequences. Those consequences might well include a greater public understanding of the natural or social world.

In contrast, Daniels (1967) has argued that the perceived distinction between basic and applied research was developed in an effort to create the conditions of autonomy in American science. By creating the rubric of basic research, researchers could justify activities that seemed bizarre and even outrageous (e.g., breeding rats) to a poorly educated public. They could and did argue that by permitting them a high level of autonomy in their work, they would contribute to the public good in the long run.

Others might argue that applied or finalized (Schafer 1983) research is less autonomous than basic or fundamental research. Yet, as Pierre Bourdieu (1997) pointed out, both basic and applied research require a high degree of autonomy. There are two reasons for this: On the one hand, without autonomy even applied researchers will be forced to look for answers to concrete problems from within too narrow a range of options. On the other hand, the very distinction between basic and applied research has of late disintegrated. Especially with respect to molecular biology, what is considered a breakthrough in basic research one day may well be a tool or product the next. Indeed, a central aspect of molecular biology is the difficulty in distinguishing between research initially of primarily scholarly interest and research likely to lead immediately to new products and processes.

Without substantial autonomy, scholarly work is likely to fail in its objectives. It becomes subject to the political or economic whims of the moment; critical issues are ignored or papered over. But autonomy does not come easily. Throughout the history of American universities, there have been those who have wished to reduce the autonomy of the academic enterprise. The agricultural and related sciences have been particularly vulnerable to such attacks for more than a century. Half a century ago, Charles Hardin (1955, 86) noted sadly, "The writer knows of no college of agriculture in which some professor has not been subjected to pressure—attempts to get him fired, to silence him on an issue, to force retraction of a publication, to require that a controversial manuscript be reviewed by representatives of an affected interest, or simply protest enough so that he will think twice before he repeats the 'offense.'"

One of the more infamous incidents forced the resignation of T. W. Schultz from the Economics Department at Iowa State University over a paper he wrote concerning the substitutability of margarine for butter. The case had a happy ending for Schultz, who joined the University of Chicago and went on to earn a Nobel Prize. But the university's reputation was sullied for many years (Hardin 1955). In recent years the number of such publicized decisions has declined, perhaps as a result of the decline of the farm bloc itself. But documented incidents are no doubt complemented by others in which faculty members quietly yielded their autonomy in the face of administrative or clientele pressures, or in which faculty have avoided certain topics because they were likely to disturb the powers that be. As we shall see below, in the case reported here it appears that at least one faculty member's career was threatened as a result of his role as a critic of the Novartis agreement. As President Eisenhower indicated in his famous farewell address, quoted in the epigraph to this volume, protecting autonomy requires eternal vigilance (Bella 1985).

Bourdieu made another relevant point with respect to autonomy. As he put it, "the more one is autonomous, the more one has a chance to employ the specific scientific or literary authority that permits one to speak outside one's field with a certain symbolic efficacy" (Bourdieu 1997, 65; our translation). Using the infamous Dreyfus case as an example, Bourdieu saw Émile Zola's critical intervention as linked to his autonomy as an author. While such instances are no doubt rare, Bourdieu's point is not to be dismissed lightly. Established scholars in the United States and elsewhere have often been able to challenge the prevailing wisdom in ways that irritated the authorities but upheld other publicly recognized values (e.g., U.S. faculty activism against the Vietnam War, physicist Andrei D. Sakharov's fight for human rights in the Soviet Union, Vandana Shiva's criticism of genetically engineered crops).

Diversity

Creativity is often constrained by a lack of diversity. The goods that universities provide are nurtured and made more robust through diversity. A diversity of standpoints is essential to the debate and dialogue that must surely be central to a great university. In its principles of community, the University of California notes, "We recognize the intrinsic relationship between diversity and excellence in all our endeavors" (University of California, Berkeley 2005). Similarly, the 2002 strategic plan notes that "social and cultural diversity are essential to the university. They stimulate creative thought and new paths of inquiry, ensure that the research questions we tackle address the whole of society, and enable us to train leaders who encompass the entire spectrum of Californians" (UCB Strategic Planning Committee 2002, 2).

Even if one accepts the somewhat problematic notion that the study of nature provides us with clear answers to problems posed, we are the ones who pose the problems. The same is true of measurement: "All measurements depend on embodied choices of apparatus, conditions for defining and including some variables and excluding others, and historical practices of interpretation" (Haraway 1997, 116). We (i.e., both the scholarly community and the public at large) are the ones who interpret the results in light of our cultural values, long-standing traditions, metaphors,[1] culturally defined meanings, and endlessly repeated stories.

For example, Donna Haraway (1997) notes that accounts of the human genome project have been given by supporters as a story of human salvation. In contrast, some detractors have argued that it may involve the end of our humanity (Fukuyama 2002). Similarly, the story of agricultural biotechnology is told by supporters as a tale of the end of famine and hunger, while detractors describe it as the beginning of a horror story, the advent of frankenfoods. As we shall see, there are also many stories of the agreement between the Novartis Agricultural Discovery Institute and the University of California, Berkeley, and each appeals for justification to one form or another of, and different routes to, creativity, autonomy, and diversity.

This is not to suggest that anyone necessarily gets the story wrong, but that there is no one, true story that is independent of human knowers in whom we can have (more or less) confidence. This includes the story that we tell in this volume. Given that aspect of the human condition, diversity is one inadequate and partial, but necessary, remedy to the limits of human beings, human society, and human culture.

Traditionally, in the Western world in general and in the United States in particular, scholarship was the exclusive province of white, Christian, heterosexual, upper-class males. While most scholars today would agree that these narrow criteria are far too confining, one cannot just, for example, "add women and stir." Such an approach ignores the fact that modern scholarship (especially in the natural sciences, but also in nearly all other fields) began as a quintessentially white male endeavor (Merchant 1982). Thus the epistemological tradition itself was based on certain assumptions about the range of issues to be addressed, the kinds of methods to be employed, and even how results were to be presented to others.

Such assumptions stand out in relief when one looks at them in historical perspective. Francis Bacon contrasted his new masculine science with old wives' tales that were to be dismissed (Bacon 1994 [1620]; Merchant 1982). Henry Oldenburg, an early member of the British Royal Society, specifically called for the construction of a "masculine philosophy" (Keller 1985). In a more detailed analysis, Shapin and Shaffer (1985) contrast the experimentalism of Robert Boyle with the mathematical approach of Thomas Hobbes. Boyle's experiments with the "air pump" (what we would call a vacuum pump today) required "modest witnesses," properly educated gentlemen who would observe the experiments conducted and draw conclusions from them. Hobbes and many of his contemporaries challenged Boyle and his colleagues, arguing that there was no way to know if the air pump generated a vacuum or something else. Thus, for them, experimental knowledge was and would always remain flawed, while the theorems of mathematics were and would always remain true by definition. What Hobbes and his contemporaries did not challenge was Boyle's assumption that science was a male enterprise.[2]

Moreover, the very notion that objectivity resides in individuals, a bulwark of scientific tradition since the days of Galileo, Bacon, and Descartes, is itself problematic (cf. Hull 1988). It presumes that individuals can transcend their location in the world. While this may be true to some degree, it is quite clearly the case that each of us has a standpoint marked by time and space, culture, gender, class, and status, among other things. Thus each of us has a necessarily partial view (Harding 1991). By enhancing the diversity of perspectives one can produce what Harding calls strong objectivity, i.e., objectivity that resides in the organization or the community rather than in individuals, and that calls background beliefs into question in making knowledge claims. Such an approach avoids both the Scylla of absolutism and the Charybdis of relativism. It makes claims neither to absolute knowledge, the unquestioned superiority of any one standpoint, nor to pure relativism, the equality of all possible standpoints. Some standpoints have more credibility than others not because they are more factual but because they are institutionalized in more complex networks.

Together, these three principles—creativity, autonomy, and diversity—distilled from analysis of texts and interviews, bring into focus what higher education ideally is and how it contributes to the common good. Together these three principles enable the university to generate knowledge, inventions, and innovations, to translate and disseminate knowledge in ways that foster the growth and development of people and communities, and to contribute to the wide-ranging discourses on ecological, social, medical, cultural, and developmental issues of contemporary politics. They are the standards by which universities should be measured. Of course, as Michael Walzer has reminded us, "There is no single standard. But there are standards (roughly knowable even when they are also controversial) for every social good and every distributive sphere in every particular society; and these standards are often violated, the goods usurped, the spheres invaded, by powerful men and women" (Walzer 1983, 10). This book is an account of a set of events that many stakeholders perceived as one such violation, usurpation, and invasion.

Part of the reason that this study is grounded in, but not an extension of, conventions theory lies in the historic contestation over the standards by which the principles of creativity, autonomy, and diversity are to be judged. Both of the foundational goals of modern universities—the Kantian pursuit of an educated and capable citizenry and the Humboldtian pursuit of excellence in scientific development—necessitate creativity, autonomy, and diversity. At the same time, the means by which these goals are pursued, and the worlds of justification to which defenders of each tradition appeal, are notably different. In the context of the development of U.S. universities in the twentieth century, with the explosion and diversification of public higher education, not only have the natural and social sciences generally won out over the humanities, and not only has professionalism won out over citizenship, but the justifications for and means of achieving technoscientific creativity, autonomy, and diversity have evolved. In various places and times, schools have pursued different (institutional and justificatory) strategies with respect to their relation to the transformation (or reproduction) of domestic, market, industrial, civic, inspirational, and opinion worlds.

As we shall see, the agreement between UCB and NADI both promoted and challenged these three principles. This was the case not solely for the research that was the subject of the agreement, but with respect to the educational role of the university as well. This agreement became an icon for these larger issues.

TESTS AND TRIALS

As noted above, proponents of CT also argue that in addition to its own largely distinct principles of justice, each world also has tests and trials for determining the greatness of both people and things. As mentioned earlier, to properly judge someone or something, one must know the corresponding principles of justice (of worth) involved. Both universities and the persons who inhabit and construct them must go through various trials (in the social, legal, and scientific senses) in order to demonstrate their greatness. Faculty members must pass through the trials of tenure and promotion; universities must be accredited by their regional accreditation associations. Universities are subject to endless public criticism and review both from within and from external constituencies. Such criticisms range from those that spring from conventions well established within the university to those that originate elsewhere and are seen as conflicting or even incommensurable.

For example, universities, such as UCB, have been frequently challenged with respect to academic freedom. Great universities have withstood attacks by external critics, arguing successfully that "Certainly science and probably every other study in the university is more successful, judged in purely academic terms, when it is free from either political control or the dominion of commerce" (Dworkin 1996). Similarly, despite intense external pressures, some administrators of great universities have supported faculty whose views they personally found abhorrent or wrong.

Universities also develop tests and trials for those who inhabit them. In an ideal world of scholarship, individual faculty members are rewarded for their support and furtherance of creativity, autonomy, and diversity. They are rewarded for publishing their research or scholarship in scholarly journals, for inventing new machines or processes that serve the public good, for generating excitement, interest, and critical thinking among their students, for taking part in public intellectual life. To do these things requires recognition generated outside the university (or at least outside any particular university) by the discipline or field (Bourdieu 1997). Furthermore, the rules for demonstration and refutation vary from field to field, yet if they are to continue to be effective, they must remain largely autonomous, beyond the control of other organizations. For example, decisions in plant biology to use *Arabidopsis* as a model plant, to determine that certain tests are the best available for determining genetic structures of plants, and even to adhere to Darwinian frameworks, are and must be independent of any person or organization outside the discipline.

Of particular import is that these tests and trials are relationships between individuals and those around them. As Walzer (1983, 3) argues in his pathbreaking work *Spheres of Justice,* "My place in the economy, my standing in the political order, my reputation among my fellows, my material holdings: all these come to me from other men and women." The same holds true for those positive and negative sanctions imposed by universities. However, not all rewards are distributed in the same way, either within universities or across all organizations.

Goods may be distributed based on market exchange, need, or dessert (Walzer 1983). Universities distribute their goods in all three ways. Who determines which rules for distribution shall apply is obviously a critical question. Much formal education and research may (only) be purchased through tuition and through research grants and contracts, respectively. Thus some sort of competitive sphere is in operation, though not necessarily a free market or even a highly competitive one. In most public universities tuition covers only a fraction of total costs, with state funding, gifts, and grants covering most of the costs. Some students may have their tuition reduced as a result of need, while some faculty will provide consulting services at nominal or no cost in cases of need. And universities are particularly generous with respect to desserts—diplomas, honorary degrees, certificates of commendation, and awards of all kinds. Honorary sculptures and even monuments are commonplace sights on university campuses. Universities may also provide funds to faculty to support research and education on the basis of some perceived need or dessert. Universities also dispense negative desserts, such as denials of tenure, expulsions, and dismissals, for, say, engaging in sexual harassment, or for plagiarism.

Of course, these three categories are hardly watertight. Frequently buildings are named not based on dessert but on the fact that funds for their construction were provided by a particular (usually wealthy) benefactor. Sometimes the children of alumni are admitted to the university simply because their parents attended the institution and not because they either needed or deserved to be admitted. Less frequently, some persons are the recipients of undeserved honors because they have political connections. And sometimes those who have transgressed university rules of conduct or even violated legal norms fail to get their just desserts for reasons of need or market exchange.

Moreover, some would attempt to impose on all university activities a single means for distributing goods. Currently it is proponents of the market who are most likely to insist that a particular means for distributing social goods should trump all others. But if university education and research are in fact needed to serve some public purpose, then they must be distributed in some manner other than solely via the market. As Walzer (1983, 89) puts it, "what we do when we declare this or that good to be a needed good is to block or constrain its free exchange." Thus a critical question for (public) universities is what roles the market, need, and dessert should play in gaining admission to the university, in partaking of its many goods, in rewarding students, faculty, and staff, and in its overall governance.

Through the many tests and trials to which universities are subjected, they rise and fall in the partially ordered hierarchy of the world of universities. The world of the university is partially ordered precisely because it has multiple and even conflicting goals. The large research universities are tested with respect to the quality of their teaching, research, and outreach, the size and scope of their libraries, the quality of their faculties and students, the rankings of departments, schools, and institutes that they comprise (by numerous ranking organizations, ranging from the National Academy of Sciences to *US News and World Report*), both the organizations that accredit the universities as a whole and those that accredit programs in various fields. And, for better or worse, they are even ranked by the quality of their football or (men's) basketball teams. The question "Which university is best?" implies that one particular portion of the hierarchy is of interest, and is an appeal to a singular world of justification. Clearly, no general answer to or single justificatory terrain for answering such a question can reasonably be achieved.

The same applies to individual students, faculty, and staff. They too are subjected to a series of tests and trials. They too rise and fall in a partially ordered hierarchy inside the university. Thus students are frequently tested to determine what and how well they have learned. Faculty are granted salary increases and promotions based on their record of teaching, research, and outreach. Staff are judged on the basis of their performance in their respective jobs. And here, too, one can rank students, faculty, and staff only in a partial manner. No universal hierarchy of greatness can be said to exist.

MULTIPLE WORLDS

It should be noted that all of us live in multiple worlds, worlds that collide, transgress upon one another, pose contradictions. One cannot be *only* a parent, an industrialist, a salesperson, a teacher, or a basketball player without abandoning one's other duties and disappointing the expectations of others. For example, what is necessary to be a good merchant may sometimes thwart what is needed in a good parent. A faculty member might feel pressured into raising the grade of an athlete, or a university administrator might feel pressured to admit a student with inadequate preparation because that student's parents are major donors. In some cases these issues arise because someone or some group is acting in bad faith or deceitfully. But in many, if not most, instances, these contradictions occur because those involved are striving to achieve contradictory or simply incompatible notions of the common good. They are attempting to grapple with the overlapping social worlds embodied in almost all lives and institutions. All of us are faced with these trials and must navigate our way through them by negotiation (e.g., Strauss 1978) and compromise (e.g., Benjamin 1990).

In fact, in a somewhat paradoxical way, attempting to focus one's life on a single world to the detriment of all others may result in one's being downgraded in the eyes of one's compatriots. For example, the outstanding teacher or researcher who ignores his family obligations may lose some of the prestige that might otherwise be accorded to that attainment. One is likely to say of such a person that "*despite* his outstanding teaching and research, he ignores his wife and children."

Moreover, it is not only impossible to live in a singular world; different worlds often collide. Such collisions may result in a conflict that is resolved either by one world trumping another or by some form of compromise. The machinist who asks the boss to hire her niece may succeed or fail (the world of industry or the domestic world may trump the other) or it may result in compromise (the niece is hired as an apprentice and is expected to learn the job rapidly). As we shall see, such conflicts are commonplace in higher education, and each challenges its common higher principles. Of particular relevance for this study are the conflicts between the world of higher education and those of industry and the market.

Yet another aspect of conventions theory is its recognition that even within a given world, not everyone shares the same goals. (In contrast, both Parsonian sociology and the theory of the firm in neoclassical economics presume that everyone within a given organization shares or ought to share the same goals and objectives.) CT emphasizes that within a given organization some people may be firmly committed to the goals of the organization, while others see the organization as a means to improve their standing in another world. For example, not all faculty members share the common higher principles surrounding teaching, research, and service found at large research universities. Some may be committed to only one or two of these domains, while others might reject all three, perhaps because they are overwhelmed by events in their domestic world, or perhaps because they have lost enthusiasm for their chosen profession. The same is equally or perhaps more true of those who are not faculty members. Administrative assistants, janitorial staff, accountants, and others employed by universities may well have their commitments elsewhere. A given assistant or accountant may work at a university because it pays the bills or because of the complex challenges the job affords.

In each of these cases, there are conventions that govern the work to be done. In the more routine jobs, such conventions may restrict the amount of work that any one person accomplishes to some amount conventionally seen as acceptable. Doing too much of such work may create resentment among one's colleagues; doing too little may bring down the wrath of a supervisor. Neophytes learn quickly what is acceptable and what is too much or too little. In less routine jobs, too, there are conventions. While faculty usually have considerable autonomy in their jobs, they too are subject to conventions. The outstanding faculty member may be praised for the quality of her or his teaching or research, but may also arouse a certain jealousy in colleagues. Similarly, the faculty member who fails to achieve even minimally acceptable evaluations from students is likely to be subjected to some sort of disciplinary action.

The issues with respect to the University of California, Berkeley–Novartis agreement (UCB-N) that we have learned about in our research and developed in our analysis all revolve around the contestation over the institutional meaning, weighting, and forms of creativity, autonomy, diversity, exchange, need, and dessert. At UCB in general, and perhaps even more intensely within the College of Natural Resources (CNR) in particular, multiple worlds have collided, transgressed, and contradicted one another in the events leading up to the agreement and in the controversy that has ensued. We noted above that Sandra Harding has argued that strong objectivity lies in the community rather than in the individual. Part of the controversy over UCB-N lies in unaddressed, much less resolved perspectives on the relative importance of individual versus communal diversity, autonomy, and creativity and the public versus private character of the modes of exchange, the definition of those in need, and the structure of desserts to be elaborated within the university. Our review of the agreement, of its history, controversy, and consequences, will seek to make this clear.

TWO

The Changing World of Universities

NIVERSITIES ARE NOT isolated institutions. Their existence and
success depends on support from various political, economic, and
nonprofit institutions. Thus, to understand the university, it is necessary to examine its role in society and its relations with other leading institutions. While UCB-N and the ensuing controversy are deeply embedded in
the highly specific conditions of the history of UCB, the Bay Area, and the state
of California, they also reflect a wider set of transformations in the national
and international character of universities, and the associated debates over
those changes. In this chapter we situate UCB-N in the context of contemporary political and economic conditions of universities, universities' response
to these changes, and various interpretations of the meaning and consequences
of these changes.

The first part of this chapter outlines the changing political and economic
conditions of universities. Beginning with the birth of the modern research university during World War II, we discuss shifts in government policy, university
funding, and economic development. The second part of the chapter examines the response by universities to these changes as presented by leading contemporary theories of universities and university-government-industry relations. The third and final section examines changes in university culture and
missions resulting from the state of modern research universities and the consequences for the future of "the university," "the multiversity," or whatever the
most appropriate future term may be.

UNIVERSITIES IN THE TWENTIETH CENTURY

By comparison to the present, during the first half of the twentieth century there was little government funding of university research and—as a result— few Humboldtian commitments to linking teaching and research held sway (Delanty 2001a, 52), particularly at public universities. Along these lines, in the era before the GI Bill, the cold war, and the space race, the primary role of universities—including the land grant, agricultural, and historically black universities—was the training and reproduction of social elites, and most university research was tied to this purpose. With the advent of World War II, the role of research in the university began to change, as many leading universities became key sites for massive, high-cost, military scientific and technological development. Indeed, the best-known example of this was the collaboration between a number of elite universities in the Manhattan Project.

The successful pursuit of large military research grants enabled universities to construct new and productive research laboratories and capacities. The need for the subsequent reproduction of these infrastructures and faculties was crucial in the establishment of the postwar treadmill of government funding for university research. The effect was to foster the aggressive promotion of a new role for universities—scientific research for the public good—a program and set of goods that were to be largely government funded.

Following the war, leading scientists and university presidents made a concerted effort to procure continuing funding of research. One of the leaders of this effort was Vannevar Bush. In *Science: The Endless Frontier* (1945), Bush laid out a scientific and technological division of labor by which university scientists would generate "basic" research and industry would turn that research into useful applications and products. Government granting programs would vet research proposals by means of a peer-review process that would ensure fiscal responsibility and scientific excellence. While never fully implemented, Bush's program created the legitimating rhetoric necessary to justify massive government spending in the creation of a new federal bureaucracy and national scientific agenda (Sarewitz 1996). With the support of industry and the military, Bush's report sparked dramatic changes in the nature and scope of U.S. government support for science. The result was the birth of the American form of the modern research university.

After the war, the Department of Defense expanded its wartime research program and became a leading source of federal research funds. Additionally, the National Institutes of Health (NIH) were reorganized into a research agency with expanded internal research capabilities and an office to oversee

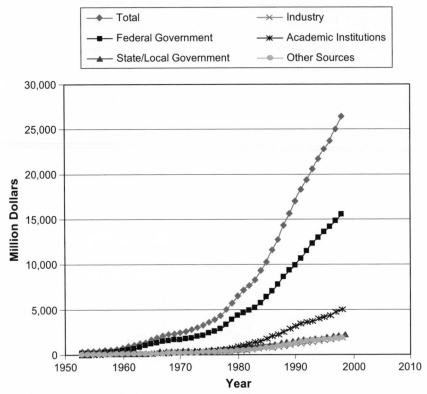

FIGURE 2.1   Sources of research support (constant 1992 dollars).
Source: Adapted from Greenberg (2001, 480).

the funding of external research. Perhaps most important, the National Science Foundation (NSF) was also established following the passage of the National Science Foundation Act of 1950. Originally proposed by Bush, it was designed to disburse government funds to keep America at the forefront of scientific research. The NSF differed from the Department of Defense and NIH in that it was only to fund "basic" research and related educational activities. Other federal agencies, such as the Department of Energy, also developed extramural grant programs. The result in the decades after the war was an effectively logarithmic growth in federal spending for academic scientific research (Figure 2.1).

The 1950s and 1960s are often referred to as the "golden era" for U.S. universities and scientific research. During this time there was continual expansion of federal government funding for research. The budgets of the NSF and NIH both expanded significantly, as did the research budgets of the Department of Defense and Department of Energy. State governments also made massive investments in universities and university systems as they sought to strengthen and expand their public universities. In this context, the UC system experienced tremendous growth, expanding from three to nine campuses between 1959 and 1966.

But education and scientific research were just one of the areas in which the U.S. government was investing heavily throughout the 1950s and 1960s. The government was also funding the expansion of infrastructure (e.g., highways), the social welfare system (e.g., unemployment and healthcare), and various military campaigns (e.g., the Korean and Vietnam Wars), among other things. The result was a significant increase in state expenditures. As long as the economy continued to grow and tax revenue increased accordingly, this was not a problem. In the early 1970s, however, the economy slowed, oil shocks hit, and stagflation was the result. In this context, falling income, business and property tax revenues, rising federal expenditures, and a nascent tax revolt produced conditions of fiscal crisis at federal, state, and municipal levels (O'Connor 1973). The response, initiated at the end of the Carter administration and more dynamically during the Reagan years, was a major restructuring of the economy (in the form of greater concerns with flexibility, information technologies, and international markets) and the state (in the form of programs to reduce the regulatory and tax burden of the state on business and increase the disciplinary character of welfare and police activity) (Harvey 1989; Jessop 2002).

The most immediate impact of these changes on scientific activity within the university was a decline in the rate of increase of government funding for scientific research. Following on the heels of a decade of double-digit inflation, the effect of this relative decline was a reduction in real educational expenditures in research and per student across the academy as a whole (Geiger 2004). Consequently, many universities cut back on many of their programmatic commitments, offered early retirement packages to senior faculty and staff, and instituted hiring freezes to reduce their operating expenses. This marked the end of the so-called golden era. From this point onward, funding and the threat of cutbacks would be a central focus of administrators, especially at public universities, and a continual part of the politics of public science.

At the same time that government funding of university science was becoming more constrained, the federal government was placing more regulations and requirements on universities, which increased their costs. For example, complying with affirmative action policies and the development of student aid programs led to increased administrative costs (Geiger 2004). The growth of universities, and the concurrent increase in PhDs, also meant that more universities and scientists were competing for funding (Greenberg 2001).

In addition, biological and physical research was becoming notably more expensive as equipment became more complex. This situation continues today.

Concerned that Europe and Japan had caught or surpassed the United States in the development of profitable technologies, the U.S. government sought to stimulate the development of civilian and consumer technologies legislatively by (a) enhancing the transfer of technology from universities and federal research laboratories to industry, (b) making changes to intellectual property laws, and (c) promoting university-government-industry collaborations (Slaughter 2004). While not realized at the time, these policies would profoundly affect approaches to and the legitimation of autonomy, creativity, and diversity in university-based scientific research.

Two changes in intellectual property rights law merit particular attention. The first is the Bayh-Dole Act of 1980 (*U.S. Code* 35 [1980], §§ 200-212) which extended patenting and licensing privileges to inventors in universities and government laboratories. Specifically, the Bayh-Dole Act "explicitly recognized technology transfer to the private sector as a desirable outcome of federally financed research, and endorsed the principle that exclusive licensing of publicly funded technology was sometimes necessary to achieve that objective" (Jaffe 2000, 535). Together with the Stevenson-Wydler Technology Innovation Act (*U.S. Code* 15 [1980], §§ 3701-14), which made technology transfer part of the mission of all federal research laboratories, Bayh-Dole marks the beginning of a shift regarding how best to translate "basic" scientific research into useful technologies and products. These two pieces of legislation not only made it possible for universities to expand techniques to patent and license research developed by faculty, to collaborate with industry, and to hold equity in startup funds. They also fostered an entrepreneurial orientation among university faculty—particularly within the fields of computer engineering and biotechnology—generating sometimes an exodus of faculty and sometimes a scientific revolving door between universities and industry.

The second change in intellectual property rights is the extension of patent rights to new areas, particularly life forms, software, and genetic research technologies (Jaffe 2000). In 1980 the Supreme Court ruled in *Diamond v. Chakrabarty* (447 U.S. 303) that life forms—as opposed to the processes by which an organism was produced or for which it was used—could be patented if they were made by humans. Although the case involved a microorganism, it was soon applied to all life forms. As a consequence, and particularly in the context of the fiscal constraints imposed in public universities during the 1980s and early 1990s, medical and agricultural biotechnology research became more directly marketable, as intellectual patenting and commercial licensing opportunities grew. At the same time, companies in these new industries became more interested in university research, and the interests of many university administrators and scientists converged with those of specific industrial firms.

The possibilities for universities created by these legal and policy changes in federal scientific programs expanded slowly during the 1980s but blossomed in the 1990s, when public universities again faced significant reductions in government funding. These reductions were related to the end of the cold war and the rapid decline in military research investments (Slaughter 2004). Furthermore, an economic recession in the United States resulted in major cuts to public university budgets nationwide (Barrow 1996). From 1990 to 1994, state budget expenditures on higher education decreased at an annual rate of 2.1 percent in constant dollars (Heller 2001). In California, which was hit particularly hard by the recession, the state's contribution to UC was reduced by 20 percent between 1990 and 1993. At UCB this resulted in the loss of large numbers of faculty positions to early retirement and unfilled vacancies, substantial increases in tuition, and major cost-cutting restructurings (Geiger 2004). At the same time, the President's Council of Advisors on Science and Technology noted that "it is unreasonable to expect that the system of research-intensive universities will continue to grow as it did during periods in the 1960s and 1980s" (Barrow 1996, 456).

With government funding becoming more constrained and less stable, university administrators began to seek alternative sources of funding and developing new ways of generating revenue. To compensate for shortages in government funds in support of teaching and overhead, universities have turned to industry as one alternative source of research funding, funding from which indirect costs could be taken and redirected to areas less well funded by state legislatures. University administrators now aggressively encourage faculty to seek industry funding and to actively seek out industry partnerships. Universities are also increasingly participating in the market. Efforts to generate revenue through patents, licenses, and spin-off companies have become standard practice for most research universities. Associated with this has been a dramatic increase in the number of patents issued to universities, despite the fact that few generate meaningful revenue streams. Furthermore, to facilitate collaboration with industry and participation in the market, universities have expanded and reorganized their administrative structures. Technology transfer and intellectual property rights offices are now common at large research universities.

While government funding of universities improved in the mid- to late 1990s, the focus and character of that funding had changed. At the federal level, as research priorities shifted from defense and nuclear development to bio- and information technologies, more money was being funneled through the NIH and less through the Departments of Energy and Defense.[1] This led to a shift in the distribution of funding among university departments. At the state level, while direct funding to universities through legislative appropriations continued, state governments also began to develop alternative means of funding universities. For example, states began to set up collaborative programs with industry, where funding for particular research projects would be jointly funded by government and industry.

These changes have produced a world much different from the one universities occupied during the postwar era. Increasingly, university research does not fit the ideal linear model first outlined by Vannevar Bush, in which government funds "basic" research, university scientists conduct "basic" research, and industry turns this research into useful things (Etzkowitz 1998; Slaughter 2004). For Bush, and for many senior faculty members today, the autonomy and creativity of universities and university science—particularly in its basic form—depended on the separation of the academic and scientific realms from the political and economic spheres. Yet today, how research is funded, conducted, and translated into useful applications and products has become much more of a hybrid, complex, and diversified quasi-public/quasi-private phenomena. Consequently, the divides between basic and applied science, technological research and development, public and private spheres, and political and scientific interests have become increasingly blurred.

In response to these challenges, and often with little reflection, universities are developing new institutional models and new societal roles for scientific knowledge. For example, Derek Bok (2003), former president of Harvard, has argued that universities are actively pursuing these new ways of operating without sufficient—much less public or democratic—consideration of the potential long-term implications for the future of universities. One potential likely consequence is an unintentional limitation on the range and diversity of academic research agendas. As is made quite clear by the UCB-N case, many of the changes now taking place at universities are quite controversial and deeply contested. It is to the theorization of these new models of operating and new societal roles—and their consequences—that we now turn.

THEORIES OF THE UNIVERSITY

Several theoretical frameworks have been developed to explain the way the changes discussed above are affecting universities. These include "private-interest science" vs. science in the public interest (Krimsky 2003), "academic capitalism" (Slaughter and Leslie 1997; Slaughter 2004), "the triple helix" (Etzkowitz and Leydesdorff 1997, 1998; Leydesdorff and Etzkowitz 2001), public-private "isomorphism" (Hackett 1990), and "asymmetrical convergence"

(Kleinman and Vallas 2001). Each theorizes the changing relationship between universities, industry, and the market, and the implications for research, teaching, and the university's role in society.

Private-Interest Science

Krimsky (2003) focuses on the social and scientific consequences of the fiscal and economic changes in higher education and scientific research. First, he argues that in order to offset reductions in government funding, universities have opened themselves broadly to industry. Not only are universities making it easier for industry to collaborate with them, or with specific faculty; universities are also actively pursuing industry-research partnerships, thereby excluding programs not easily supported by industry funds. Second, Krimsky notes that administrators are managing universities more and more as if they were for-profit businesses. Administrators are encouraging scientists to take an entrepreneurial approach toward their research—an approach that may or may not be oriented to private sources of research funding and collaboration—as part of a generally more corporate approach to university management. Krimsky argues that this is evident in the emphasis now placed on the potential monetary gains of research through such things as patents and spin-off companies, as well as the kinds of audits and success/promotion measures brought to bear on departments and faculty.

For Krimsky, these changes represent the commercialization of the university. Whereas scientific knowledge used to be produced for its intrinsic value as a public good, with the commercialization of the university, scientific knowledge moves increasingly toward instrumental values and the private good. In other words, the range of research problems and the methodological means by which they are approached are increasingly driven by questions of fiscal reward to universities and profitability for corporate partners. Krimsky argues that these trajectories are altering the overarching mission of universities. As he sees it, the purpose of the university is now "to provide the personnel, the knowledge, and the technology for economic and industrial development" (177) as opposed to the historical commitment to the generation of public knowledge, an informed and critical citizenry, and the search for solutions to major social and environmental problems.

Academic Capitalism

Similarly, Slaughter, Leslie, and Rhoades (Slaughter and Leslie 1997; Slaughter and Rhoades 1996; Slaughter 2004) argue that universities have shifted from a public-good model to an "academic capitalism" model. In their 1997 book, Slaughter and Leslie stressed that universities had begun to engage in "marketlike efforts" as a way to offset reductions in government funding. More recently, in 2004, Slaughter has added an emphasis on the rise of the "new economy," rooted in knowledge-based industries (from biotech to infotech), to her earlier concerns with the role of budget shortfalls in the rise of academic capitalism.

Specifically, these researchers define academic capitalism as (1) the pursuit of profits by market means, including patenting and licensing faculty research, spinning off companies from university enterprises, and close financial ties to corporations, and (2) the development of public-private knowledge networks in which universities and industries collaborate closely. Whereas academic capitalism remained an emergent phenomenon at the time of their original formulation, Slaughter and Leslie argue that it is now deeply embedded in the structure of universities. In other words, managing universities as if they were for-profit businesses has become the norm.

As procuring money by means of research funding and the commercialization of results become more and more important, the reward structure of universities is subtly changing, so that research that is closer to the market is more highly rewarded. As evidence, Slaughter and Leslie (1997) note the rising centrality of technoscientific fields and in particular the biological and medical fields, which are closer to the market and tend to have more resources. They contrast this to the social sciences and humanities, which tend to generate neither large grants nor commercializable products and are experiencing an accelerating decline in resources and support. Extending Krimsky's argument, they suggest a likely narrowing of the kinds of research undertaken within technoscientific disciplines and across universities as well.

The greatest danger posed by academic capitalism may be an erosion of the privileged position that universities hold as neutral arbiters, objective witnesses, and fair-minded teachers in society. Historically, universities have had an aura of separateness, relatively unconstrained by economic and political interests and above the passionate fray of social struggle.[2] As universities and their scientific and technological developments come to resemble businesses more and more, Slaughter and Leslie worry that they will be treated more and more like private firms and that faculty will be treated more like business professionals. An indication that this in fact may be occurring is the recent ruling by the U.S. Court of Appeals for the Federal Circuit in the case of *Madey v. Duke University* (307 F.3d 1351 [Fed. Cir. 2002]). In that case, the court found that Duke University may not claim a noncommercial experimental-use exemption for the use of patented instruments. The court noted that rather than using the instruments "solely for amusement, to satisfy idle curiosity, or for strictly philosophical inquiry," they were being used in furtherance of the university's business objectives as a research institution. As a consequence, depending on how the decision is enforced, universities may have to pay royalties and licensing fees if their faculty use certain patented technologies.

Slaughter and Rhoades (1996, 330) note that that universities "want the best of both worlds—the protection and continued subsidies of the public sector, and the flexibility, opportunities, and potential revenue streams of the private sector." However, universities run the risk of jeopardizing continued public support as their practices increasingly resemble those of private enterprises. For example, state governments may see little benefit in funding an institution that is primarily working with and for industry in its research arm, or is generating its own revenue through direct participation in the economy. As universities and university research tend not to be especially profitable—notwithstanding the rare and extraordinary successes a few universities have had—further loss of public support could generate a serious crisis for universities (Slaughter 2004).

The Triple Helix

In contrast to Krimsky and the group of scholars with whom Slaughter has worked, Etzkowitz et al. (1998) emphasize the long historical engagement of universities with business interests and economic activities. However, they also argue that, until recently, the boundary between universities (public) and industry (private) was strictly policed and maintained. Thus, whereas previously interaction between universities and industry took the form of consultation and philanthropy, today industrial interactions are more direct and pervasive, particularly with respect to research funding and joint-venture projects.

Again in contrast to Krimsky and Slaughter, Etzkowitz and his colleagues have suggested that the rising value of scientific knowledge and technological capacity is more important than reductions in government funding when it comes to the development of the entrepreneurial university (Etzkowitz et al. 1998; Webster and Etzkowitz 1998). With the shift toward "knowledge-based" economies, fundamental technoscientific research has become more important to firms and their profitability. The effect has been the development of new kinds of interactions and relations between universities, industry, and government, interactions that do not adhere to the public-private divide that previously governed such relations. Furthermore, Etzkowitz and Leydesdorff (1997; Leydesdorff and Etzkowitz 2001) emphasize the tripartite changes in relations between universities, industries, and governments, which they have dubbed "the triple helix."

In the triple helix, universities, private firms, and governments are linked together in networks dedicated to new modes and forms of knowledge production. The role of the university is to produce knowledge, of industry, to put it to use, and of government, to provide institutional settings and infrastructural programs to foster university-industry relations (UIR). Whereas earlier modes of UIR generated relatively indirect connections between university research and industry needs, within the triple helix university research is much more closely tied to and influenced by economic opportunities and industry needs. The effect is the "capitalization of knowledge," a process wherein commercial interests, more than intellectual curiosity, increasingly drive research (Etzkowitz et al. 1998).

With the capitalization of knowledge, Etzkowitz, Webster, and Healy (1998) argue that the internal practices of universities have been transformed in several critical ways. For one thing, universities are placing greater emphasis on intellectual property and business ventures, such as spin-off companies. However, whereas Slaughter and colleagues tend to view the increasing market orientation of universities critically, Etzkowitz, Webster, and Healy tend to avoid normative judgments. In part, this may be because the two groups of scholars have different focuses and different units of analysis. Slaughter and Leslie are largely interested in the academic spaces negatively affected by academic capitalism—the faculty and fields that are marginalized under contemporary conditions. By contrast, the capitalization-of-knowledge argument is primarily concerned with how universities can survive and thrive under current political and economic conditions. For these theorists, universities might continue to function and serve society through academic capitalism in an era of constant or declining government support.

Isomorphism and Asymmetric Convergence

A fourth approach to UIR builds on the idea of isomorphism. From this perspective, universities and industries are becoming more alike, albeit unevenly. Hackett (1990) first developed this line of thought, when he argued that changes in the external relations of the university (i.e., funding relations and their role in society) would affect the internal practices of universities in the future. First, greater dependence on the private sector for resources will lead universities increasingly to resemble the private sector. Second, "increased resource dependence and other transactions with government agencies will cause universities to adopt and enforce the rules and formal rationality of government bureaucracies" (Hackett 1990, 247). If both of these things were to occur, Hackett argues that the university reward structures would become defined in ways that are increasingly similar to those of the private sector, and formal audits of faculty would become increasingly commonplace. The effect, he notes, might be the homogenization of university science and research.

Building on the idea of isomorphism, Kleinman and Vallas (2001) argue that a process of "asymmetrical convergence" is occurring between universities and industry. On the one hand, universities are adopting the codes and practices of industry. On the other hand, high-tech firms and industries are increasingly restructuring their R&D practices according to academic norms. The process is asymmetrical, however, as industry is having a greater effect on universities than universities are on industry. The result of such convergence is a blurring of "boundaries between institutional domains" that is producing "a new and precariously integrated structure of knowledge production" (Kleinman and Vallas 2001, 453).

For Hackett and Kleinman and Vallas, it is not just the structure and formal practices (i.e., tenure review) of universities that are being commercialized, but also the culture of academe. Here, industrial norms and values are infiltrating the university not just through funding but also through less direct channels, such as the exchange of personnel. Thus, even without administrative pressure or structural changes, some faculty and departments might begin to adopt business values as a result of closer ties with industry. From this perspective, understanding the influence that industry is having on universities entails going beyond questions of funding and university policy. Interactions and relations between university faculty and industry scientists must also be examined.

THEORIES OF UNIVERSITIES IN CONTEMPORARY SOCIETY

The University in the Audit Society

To varying degrees, all of the scholars reviewed thus far acknowledge the growing corporatist dynamic in the administration of university science. Michael Power, in *The Audit Society* (1997), emphasizes the vagaries, unintended consequences, and contradictions of the kinds of outcome audits that are increasingly hegemonic within corporate and public governance practices. Relatively independent of recent changes in public funding of higher education and the rising cost of capital-intensive technoscientific research, Power notes that there "is something unavoidable about audit and related practices and something irresistible about the demands for accountability and transparency which they serve, even when the consequences are perverse" (1997, xv). But Power attributes the spread of the practice to the combined effects of economic and fiscal crises over the past thirty years, and to the patterns of neoliberal economic and governmental restructuring instituted to remedy the situation.

As with the discussion above about changes in the academy, Power argues that the "audit explosion has its conditions of emergence in transformations in conceptions of administration and organization which straddle or, better, dismantle the public-private divide" (1997, 10). Particularly in the shift from regulatory government to corporatist governance, audits are the preferred form of outcome-based performance measure. Moreover, Power insists that audits are fundamentally conservative and argues that "institutionalized pressures exist for audit and inspection systems to produce comfort and reassurance, rather than critique: If this argument is true and auditing systems are primarily about reaffirming order then it will be interesting to see how auditable outcome-based performance measurement progresses in the face of system decay, especial given political rhetorics of zero tolerance for poor performance" (1997, xvii).

As we shall see in greater detail later in this volume, the UCB-N controversy appears to be an instance where the new dynamics of university-industry relations are combined with concerns about audits. Indeed, the disciplinary and departmental restructuring at Berkeley followed in large part from low performance measures generated by new audit structures. This in turn simultaneously generated conditions that some believed would correct the shortcomings of departments and that others saw as illustrative of system decay rather than the legitimate pursuit of academic excellence.

Power, a professor of accounting at the London School of Economics, is certainly not opposed to audits, yet his concern is that audits are used—intentionally or not—to transform organizations in ways that are not always positive. "[T]here are clearly many circumstances where we think that some checking and monitoring is justified and where it would be unreasonable not to learn from the experience of disappointed expectations. What we need to decide, as individuals, organizations, and societies, is how to combine checking and trusting" (1997, 2).

The difficulty with resolving this problem is that checking up on each other is not simply a matter of technical expediency. It is also a cultural issue; it is a matter of who we are and how we live. Through their very focus on that which is auditable, the academic audits prevalent today tend to exclude discussion of means for realizing creativity, autonomy, and diversity in a changing political and organizational environment.

Furthermore, Power argues that a commitment to audits as technical instruments for evaluation also embodies a commitment to audits as normative programs of accountability and control. It is to this sort of concern that Krimsky, Slaughter, and others in the UIR literature refer when they write of the management of universities as if they were for-profit businesses. The concern is that the accountability norms of outcome-based performance audits foster modes of autonomy and creativity that will constrain academic diversity in the name of academic capitalism. Here, and perhaps most important for the argument in this book, Power argues that "auditing works by virtue of actively creating the external organizational environment in which it operates" (13). The pursuit of performance measures predicated on prioritizing results that may prove commercial, then, is often and accurately perceived as deemphasizing noncommercial scientific efforts and research that may not be commercially profitable. That this process is often perceived to have consequences for the hiring preferences of universities only exacerbates concern among many scholars and researchers.

Finally, relative to the still widespread commitments researchers have to Mertonian norms in their professional work, Power shows that "Programmatic commitments to greater accountability is far from contributing to transparency and democracy. ... Rather than providing a basis for informed dialogue and discussion, audits demand that their efficacy is trusted: Furthermore, audits themselves are necessarily trusting practices because they co-opt management systems into the auditing process" (13).

Much of the concern about science in the private interest, academic capitalism, the triple helix, and isomorphic or asymmetrical convergence relates to the traditional values of institutional transparency and democracy, values undermined rather than upheld by the preference for objective measurement of performance and productivity by means of audits.

Science as Social and Cultural

While each of the frameworks discussed above is valid in that universities are undergoing significant changes, it is also important to remember that universities have never been completely autonomous institutions (Kleinman and Vallas 2001). Furthermore, as Etzkowitz, Hackett, and Kleinman suggest, ever since Kuhn's *Structure of Scientific Revolutions* (1962), historians and sociologists of science have clearly understood that the practice of science has never quite corresponded to Merton's (1973) norms of universalism, communism, disinterestedness, and organized skepticism. For that matter, most critical science studies—from the labor process perspectives of the contributors to

the journals *Science for the People* and *Science as Culture* in the 1970s, to the
sociology of scientific knowledge (Bloor 1991; Shapin and Schaffer 1985),
actor-network theory (Callon 1986, 1998; Latour 1993, 2005), and feminist
epistemology (Haraway 1991; Harding 1991; Smith 1990) over the past quarter-
century—clearly challenge the notion that science has ever been free from
economic, political, and many other social influences.[3]

Readings (1996) argues convincingly that Western universities, and even
cold war multiversities (Kerr 1963), now reside in a postnational and multicultural world where there is no longer a single, universal national culture for
which citizens might be trained any more than there is a single, universal science to practice and teach. Furthermore, the erosion of the cultural and scientific foundations of the Kantian and Humboldtian traditions has been exacerbated by the rising prestige of a competitive and increasingly international
academic market for students, scholars, and commodified knowledge. In short,
the intellectual and institutional foundations of the university lie in ruins. In
the place of universities of reason and culture—Readings's categories—the
contemporary commercialization and governance of the university is generating a university of excellence, where "excellence" equates far too easily with
auditability.

For Readings, this loss of clarity, unity, or unifying principle as to the purpose of the university is a process that primarily played itself out between
1968 and 1989. Over this period, social, cultural, and scientific crises increasingly diversified and fractured the university. Indicative of this state has been
the rise of synthetic, hybrid, and relativist academic traditions such as cultural
studies. Readings argues that cultural studies, and the culture wars, are both
cause and consequence of the collapse of coherent national cultures in the West
and, as such, of a decline in the coherence of the traditional Humboldtian
modes of legitimating public expenditures in universities. Under these conditions, the shifting orientation of university administrators toward (increasingly) global corporations, and the pursuit of the everything-and-nothing
term "excellence," have further undermined the meaning and import of
national culture as well as the craft practices and values of university science.

Readings's solution is neither a romantic return to the science of the past,
science in the public interest—as appears to be the case in Krimsky's work—
nor a wholehearted embrace of the high-tech and market-loving present. He
advises, ironically and pragmatically, that we learn to dwell in the ruins. Times
are not what they once were, but at least our illusions have been stripped away.
Power has been decentered, and this decentering makes space for creative
knowledge production and political resistance. Such an approach, however,
makes a great deal more sense in the humanities and parts of the social

sciences, where university-industry relations are weak or nonexistent, than it does in the physical sciences. Thus the kinds of autonomy, creativity, and diversity that Readings promotes are exactly what opponents of UCB-N see as disappearing.

Gerard Delanty (2001a), on the other hand, balances the penetration of the university by private economic interests—and the promotion of academic patenting and high-tech startups initiated by university faculty—with the rising influence on university curricula of multicultural groups. In short, Delanty sees the university as engaged in the (unresolvable) struggle to adjudicate the demands of the multicultural and democratic public and civic sphere in balance with the demands of global and economic private interests, in the face of the fiscal crisis of state funding for public higher education.

Delanty's focus on science and money clarifies the importance of Readings's book, *The University in Ruins,* for our study. By stressing the growing nonacademic participation in the governance of both the cultural *and* the scientific sides of universities, Delanty shows Readings's focus on the public cultural consequences of private technoscience to be too linear. While cultural studies is one alternative for the reproduction of culture, its intellectual sibling on the other side of the university is science studies. Generated at the same time as, and in conversation with, cultural studies, science studies has asked the same kinds of questions about technoscientific institutions and practice that cultural studies has asked about the Western canon.

From the perspective of science studies, the controversy over UCB-N emerges as administrators, professors, graduate students, and NGOs struggle to deal with the globalization of the economy, postmodern cultural uncertainty, recurring fiscal crises in higher education, and questions about objectivity, power, democracy, and risk. The UCB-N controversy was replete with people acting on different assumptions about the cultural, rational, critical, and scientific grounds upon which the university ought to stand. In part this is because of the unresolved tension between the older melding of Kantian and Humboldtian commitments within the "multiversity." It is also, and probably more significantly, grounded in the lack of an open debate within the university as to what the role(s) of the university should be given the political, economic, cultural, and technoscientific changes of the past thirty-five years.

Delanty and Readings agree that the goal of the conversation should not be consensus—a new equilibrium or unity. Each argues that acknowledging, but not embracing, dissensus is a necessity. What matters is that the conversation be developed and ongoing. UCB-N was developed—and the controversy exploded—in a manner uninformed by the issues raised by cultural and science studies, much less by a public exchange over the evolving role of the university. Nor has any other major research university grappled with these complex issues more effectively. Yet these are the issues that must be debated at Berkeley and elsewhere, because the global pace of change appears unlikely to slow in the near future.

Delanty, however, provides no more guidance on this front than Power does. As we see it, the core question remains how to proceed when the university is in ruins, university-industry relations are of ever-greater importance, and outcome-based performance audits continue to rise in importance as tools of governance.

CONCLUSION

Each of the theories, as it dissects dissensus, highlights different orders of worth among multiple worlds. Some voices push for new orderings or a return to a past order. Some advocate greater isolation among worlds, while others assume the need for a working compromise or consensus between worlds. Conventions theory suggests that a diversity of standpoints, as well as a plurality of justifications for both partnerships and patronage, are strongly present. Given this theoretical grounding, we examine in the following chapter the rich detail of the situated judgments that came to define the UCB-N agreement.

THREE

Land Grant Universities, Agricultural Science, and UC Berkeley

I N ADDITION TO THE BROAD issues discussed in the previous chapter, the histories of agricultural sciences generally and of UCB specifically have affected the way the UCB-N agreement was perceived and the effects that it has had. Most important are tensions between progressive and populist research orientations that have characterized the agricultural sciences since the beginning of the land grant university (LGU) system. This divide continues today, and is most clearly evident in the split between research on biotechnology and sustainable agriculture.

Populism and progressivism, both born of nineteenth-century social and economic struggles, are deeply embedded in American politics and academic life. Populism has been manifested in both right- and left-wing versions of defending "the little guy" from large and powerful economic, political, and cultural interests, whether East Coast banking, midcontinental railroad interests, or West Coast tree huggers. Progressivism, by contrast, stresses the rational, scientific, and efficient management of everything from crime to natural resources, the public purse to productive technologies. While public and land grant universities have historically been given legitimacy by populist appeals to the public interest, the means by which that interest has been served over the last century have predominantly been progressive in character. This contradiction lies at the heart of many academic treatises (McConnell 1953; Hays 1959; Hightower 1973; Kloppenburg 1988) and of struggles such as that over the UC Berkeley–Novartis agreement.

Berkeley has a long history of success in both progressive and populist research traditions. Researchers at UCB have contributed to the progressive modernization of agriculture. At the same time, some of the landmark populist studies on agriculture and labor, the impacts of agribusiness on rural communities, and the health and environmental impacts of pesticides have been produced by Berkeley faculty. Indeed, the College of Natural Resources (CNR) has long been known for its multiple, often conflicting, research orientations. However, some faculty and agricultural/environmental advocacy organizations perceive this balance between progressive and populist research to be disappearing at UCB. From their perspective, progressive forms of research, which, as represented by biotechnology, now raise large sums of money and garner considerable disciplinary prestige, are being promoted, while research on sustainable agriculture—with its dual commitment to environmental and social justice—is slowly being eliminated from UCB. Thus, for many, UCB-N was the progressive straw that broke the back of the populist camel in CNR.

A key point here is that the UCB-N controversy was in many ways less about biotechnology and university-industry relations than it was a struggle over the mode—or modes—of academic creativity, autonomy, and diversity embedded in CNR. Additionally, many faculty concerned with or opposed to the agreement focused on the negotiation process, a process they saw suborning the deep and historical university- and campus-wide commitments to shared, open, and democratic faculty governance derived from the melding of populist and progressive commitments in Kerr's multiversity.[1] In this way, the divide between faculty engaged in conventional research and those engaged in research on alternative agriculture, including actors outside the university, has figured prominently in the response to the agreement itself and the interpretation of its implications.

This chapter seeks to situate UCB-N and the ensuing controversy in the context of LGUs and agricultural science in the United States. The first section provides an overview of the establishment and development of LGUs. In the second section we discuss the long-standing tension in the agricultural sciences between progressive and populist research orientations. The third section outlines the establishment and growth of the agricultural sciences at UC. The fourth and final section discusses the shift from agriculture to natural resources at UCB, and the recent restructurings of CNR.

THE LAND GRANT UNIVERSITIES

The LGUs were established by the Morrill Act of 1862, which mandated that "at least one college" and "a minimum of forty acres of land" be set aside for the purpose by each state. Such universities were to combine the applied arts and sciences with public service to, and teaching for, the citizens of each state. LGUs would disseminate knowledge to the laboring classes, and the knowledge produced there would be verified by public disclosure so as to ensure its universal availability and validity.

The impetus for the establishment of the land grant system came from two, often conflicting segments of society. On the one hand, the land grant system was a response to populist pressures to extend higher education to broad segments of the U.S. population (National Research Council 1995). Thus LGUs would provide "useful and relevant scientific education" for the agricultural and artisanal classes (Sanders 1999). On the other hand, the establishment of the LGUs was a response to progressive demands by bankers, wealthy farmers, and editors of many of the agricultural journals for a more productive form of agriculture. Citing Liebig's research into soil nutrients in Germany, pro-science farmers and journalists argued that greater productivity in agriculture could be achieved through the application of science (Busch and Lacy 1983). To fully realize the potential of agriculture, organized research and government funding were considered necessary. Thus, since their inception, LGUs have had to negotiate between a populist and a progressive mission.

One result of the ongoing debate between populists and progressives was the passage of the Hatch Act in 1887, just six years before historian Frederick Jackson Turner proclaimed the "closing of the frontier." With the act's establishment of State Agricultural Experiment Stations (SAES), scientists at LGUs were encouraged to support agricultural development, expansion, and intensification through applied empirical research as well as through teaching. Soon after the passage of the Hatch Act, however, questions emerged as to what kind of research was most appropriate in LGUs. Should it focus on the practical needs of farmers or on the largely esoteric questions that science had traditionally investigated? In other words, should LGUs engage primarily in basic or applied research?

The question was answered in classically pragmatic fashion. Under the Hatch Act, although each SAES received a substantial sum from the federal government, each was responsible to its respective state legislature. This was a victory for supporters of applied research, as each state legislature was eager to show results to its constituents (Busch and Lacy 1983). The practical orientation of research is evident in the initial focus of SAES on such issues as crop development and the testing and analyzing of commercial feeds and fertilizers (Collier 2002). This research agenda was also a victory for wealthy farmers and others who had envisioned the land grant system as a tool by which to increase the productivity of agriculture.

The Adams Act, passed in 1906, further solidified the research component of LGUs by providing funds to be used only for "original research" (Rosenberg 1976). Eight years later, in 1914, the Smith-Lever Act formally established the extension component of LGUs. The establishment of a formal extension component freed scientists from the time and effort necessitated by their historical outreach responsibilities and allowed them to concentrate on research and teaching. This set the conditions for the development of research agendas that, over the next century, became increasingly basic as the division of labor between research, teaching, and outreach became formalized.

In recent years, however, agriculture has become an increasingly minor part of what many LGUs do. As the number of people engaged in agriculture and agriculture's contribution to the economy has declined, funding of agricultural research has become less of a priority for the U.S. government. While agricultural research was the centerpiece of federally funded research prior to World War II, since 1976 federal funds to LGUs for agriculture have remained stagnant, even as federal support for other forms of research has grown. By 1991 less than 4 percent of federal funds received by universities supported agricultural research (Fuglie et al. 1996). State funding continued to increase until the early 1990s, when it too began to stagnate. Furthermore, even within USDA, funding for block grants (often described as formula funding, based on a formula related to a state's farm population) stagnated, as funding for competitive grants programs increased. Thus, while LGUs continue to be among the most prominent of U.S. universities, there is considerable debate regarding both the continued applicability of the land grant mission and the role of agricultural research and extension.

AGRICULTURAL SCIENCE

Despite the populist character of the land grant mission, progressive science has long dominated LGU research programs. In part this is because of the historically close relationship between agricultural interest groups, state and federal governments, and many scientists, which resulted in the censorship and suppression of populist science and research at LGUs. Even in the early to mid-1900s, while scientists at LGUs began to gain limited autonomy from clientele groups, enabling them to begin to undertake "independent," more basic research, there was still considerable resistance to populist-oriented research. Such research was often considered "activist" and thus was stigmatized. For example, research focused on social welfare and on political and economic justice was not well received by some segments of the agricultural community, perhaps most notably the Farm Bureau (Busch and Lacy 1983). Consequently, many of the most outspoken social scientists were forced out of LGUs in the 1930s and ensuing decades (Hadwiger 1982).

Despite this resistance, populist research interests over time retained a foothold as part of agricultural science. On the one hand, the control of client groups over agricultural research weakened, as agricultural scientists were able to procure funds from other sources. On the other hand, the emergence of back-to-the-land movements in the 1960s, mainstream environmentalism in the 1970s, and the farm crisis of the 1980s produced new constituencies and more demand for populist-oriented research. The result, however, was not a general widening of research across the agricultural sciences, but rather the division of the agricultural sciences into either progressive or populist disciplines.[2]

Buttel (1985, 85) argues that disciplines that are primarily engaged in research associated with conventional agriculture "tend to be most accepting of prevailing institutions … for the generation, development, and diffusion of technological knowledge." These disciplines (e.g., agronomy, crops/soils, horticulture, plant pathology, and animal science) are the most central to the progressive mission of LGUs—to enhance productivity. In addition, scientists primarily concerned with increased productivity "tend to see 'externalities' or 'social problems' resulting from technological change as being irrelevant or unfortunate, and as a set of disturbances in executing the task of more efficiently utilizing agricultural resources" (Buttel 1985, 85). For example, in a classic article entitled "The Morality of Agronomy," agronomist Boysie Day (1978) wrote, "I fear that when his [the agronomist's] values come in conflict with questions of political expediency, social justice, and equity, the production ethic takes precedence. Such is the stuff of revolutions."

In contrast, the focal concern of alternative agriculture faculty is the generation of environmental, economic, and social well-being for producers, rural communities, and consumers (Crouch 1990). These scientists tend to focus on the externalities or social problems that conventional scientists often dismiss or ignore. Faculty who do research on alternative agriculture also tend to focus on agricultural practices and processes, rather than inventions, and thus are less interested in intellectual property rights. Rather, they seek to disseminate their research findings as widely as possible. Consequently, they tend to be active in alternative networks, such as farmer-to-farmer and farmer-to-consumer networks (Hassanein 2000).

Beus and Dunlap (1990, 591) argue that the divide between progressive and populist research in the agricultural sciences "represents a conflict of fundamentally divergent paradigms." Each "paradigm" holds out a different vision of what constitutes "good" agriculture and how best to achieve it, and, as such, the two are often at odds. Today the pivotal axis of these debates is between supporters of progressive biotechnology research and proponents of sustainable, agroecological and/or organic agrifood systems.

Agrifood biotechnology research has a tendency to be pursued in departments more concerned with "basic" scientific innovations that are effectively intended to be "applied" by commercial enterprises. This basic research—whether to determine the mechanisms of insect resistance, to identify the genetic basis of drought tolerance, or to increase the nutritional value of cereals—intends to address the ecologically destructive nature of pesticides, the huge costs of constructing irrigation systems in arid regions, and the socially devastating consequences of malnutrition in underdeveloped areas of the world.

In contrast, sustainable agricultural research has a tendency to dominate in departments more concerned with applied technological innovations intended to be provided effectively free of charge to farmers—farmers who generally have insufficient funds to sponsor research in their own interest. This applied research tends to pursue questions of agroecological conditions such that natural predators, soil flora and fauna, and species diversity work to reduce the impact of pests, to hold water and nutrients in the soil, and to produce more nutritious and tasty crop varieties in the process. Furthermore, a number of social scientists with similar social and agroecological commitments counter arguments by biotechnologists about the ecological, economic, and nutritive value of genetically modified crops with analyses of the social impediments to and contradictions of pursuing agricultural sustainability or solving hunger problems with biotechnology—particularly commercial biotechnology.

THE LAND GRANT MISSION AND
AGRICULTURAL SCIENCES AT UC BERKELEY

The events that led to the founding of the University of California began in 1855 under the College of California charter (UCB 2002). In 1860, trustees dedicated the College of California's Berkeley site, which would eventually become the location of the current University of California. The College of California, founded as a private institution, was to be modeled after East Coast universities such as Harvard and Yale.

In 1866, under provisions of the Morrill Act, the California legislature founded the Agricultural, Mining and Mechanical Arts College. Congruent with the land grant mandate, the college was to teach agricultural sciences, mechanical arts, and military tactics with an emphasis on a "liberal and practical education of the industrial classes" (UCB 2002). In 1868 the boards of trustees of the College of California and the Agricultural, Mining and Mechanical Arts College decided to merge the two schools to take advantage of each college's individual strengths—the College of California's large land holdings in Oakland and Berkeley and the Agricultural, Mining and Mechanical Arts College's significant funding from public sources. On March 23, 1868, the governor of California, Henry H. Haight, signed into law the Organic Act, establishing the University of California. However, while agriculture has been part of UCB from its inception, the agricultural sciences have also occupied a tenuous position there.

E. W. Hilgard, one of the most prominent early figures in UC agriculture, became the dean of the College of Agriculture at Berkeley in 1875. At the time of Hilgard's arrival, those running the university held agriculture in low regard. Agricultural groups in California, such as the Grange, also had little respect for or trust in UC. To establish agricultural research within both the university and the state, therefore, Hilgard set about building a research program oriented toward the needs of California farmers. His dual aim was to use research as a way to maintain legislative support for the College of Agriculture and to appease the Grange (Rosenberg 1976). Despite Hilgard's efforts, the UC administration continued to hold agriculture in low regard through the 1880s. The refusal of UC officials to adequately recognize the agricultural sciences led Hilgard to look to other sources for funding. Thus Hilgard was an ardent advocate of federal funding for state experiment stations (i.e., the Hatch Act). He argued that such funding would help free agricultural science from the grip of state governments (Hilgard 1882).

In the early 1900s agricultural interests and scientists were able to persuade the California legislature to expand its funding of agricultural research. One outcome was the establishment in the southern part of the state of the Citrus Experiment Station at Riverside in 1907, so as to better conduct research on agriculture there. In 1908 the University Farm School opened at Davis, where students from UCB learned the latest in agricultural methods and technology. In 1959 Davis and Riverside would both become general campuses. While all the campuses in the UC system are technically part of the LGU system, agricultural research has been predominantly located at Berkeley, Davis, and Riverside.

At Berkeley, agricultural research has been housed mostly in the College of Agriculture, now known as the College of Natural Resources. For much of CNR's existence, faculty have conducted research on agricultural production, taught courses on farming, and been active in extension efforts. UCB has long been known as one of the leading agricultural and environmental research centers in the world. It is also known for its multiple, often conflicting, research orientations.

On the one hand, UCB has had a strong commitment to increasing agricultural production and productivity (i.e., conventional agriculture). On the other hand, it also had a core group of faculty who challenged the industrial and productivist orientations and tendencies of California agriculture through their research on alternative agriculture. For example, Paul Taylor, an agricultural economist at UCB, was instrumental in encouraging the study of migratory agricultural labor in the 1920s and '30s. Similarly, Walter Goldschmidt conducted his landmark study *As You Sow* (1978 [1946]) as a graduate student at UCB. The study compared the character of two communities in California's Central Valley, one dominated by a large, effectively capitalist, agricultural concern, the other structured around a number of smaller, independent units.

In addition, under Kerr's tenure as president, some faculty in the Entomology Department criticized what they saw as the social and ecological consequences of conventional pesticide usage—and the inappropriateness of UCB faculty testing pesticides for chemical companies—and carved out the Division of Biological Control. Within the sustainable agriculture community— a community that is populist to the core (Allen and Van Dusen 1990; Allen and Sachs 1991; Allen et al. 1991; Allen 1994)—the Division of Biological Control is often seen as one of the few footholds sustainable agricultural science was able to gain within the LGU system. The faculty and graduates of the division have developed core texts and techniques in agroecological research. Their research has given the division, and CNR, great importance to the sustainable agriculture movement throughout the world.

Thus a hallmark of CNR was a dynamic tension between competing visions of the kind of agriculture that an LGU was supposed to foster. Relations between these distinct and largely incommensurable visions served CNR exceptionally well as it became internationally known for, among other things, its commitment to both progressive research that seeks to increase the productive efficiency of California agriculture *and* populist/environmentalist research on alternative agricultural practices and natural resource studies that focus more on sustainability and community than on productivity and profitability.

With the urbanization of the Bay Area and the growth of the UCB campus, a number of teaching, research, and extension activities in the agricultural sciences were transferred to the Davis and Riverside campuses. One result was that research in the agricultural sciences at Berkeley became more and more basic. For example, the UC-wide academic plan of 1961 stated that the program in agriculture at UCB "should continue to emphasize teaching and Experiment Station research in the basic physical, biological, and social sciences, taking advantage of the vast array of scientific resources on that campus to add to the pool of fundamental knowledge upon which advances in agricultural technology depend" (UC 2004). This tension between applied and basic research, and between the kind of agricultural research appropriate at Berkeley versus Davis or Riverside, persists today. In times of fiscal crisis, for example, proposals to shift faculty in the agricultural sciences, and Agricultural Experiment Station support, from Berkeley to Davis or Riverside tend to emerge (Barinaga 1994).

THE COLLEGE OF NATURAL RESOURCES
AT UC BERKELEY

The declining relevance of agricultural issues at UCB, coupled with the increasing significance of environmental issues, led to the merger of the College of Agriculture and the School of Forestry to form the College of Natural Resources in 1974. As the name implies, the focus of CNR was much broader than just agriculture sciences. The chancellor of Berkeley at the time, Albert H. Bowker, stated that "the central concern of the college will be the renewable resources of all non-urban lands of the state" (Stelljes 2004).

CNR has been restructured several times since its creation. These restructurings are important in that they have shifted the focus of research and changed the disciplinary boundaries in the college. For some faculty, the restructurings represent an effort to make agricultural research at Berkeley more progressive and more basic. Thus, not surprisingly, some faculty feel that they and the kind of research they do have been marginalized by the way that CNR is now organized. In this way, the restructuring of CNR marked the beginning of the controversy that erupted over UCB-N, supporters tending to be those who were happy with the restructuring and critics those who were unhappy.

The first restructuring of CNR that is important for understanding UCB-N and the ensuing controversy occurred in the late 1980s as part of a campus-wide reorganization of biology. At the time, some faculty and administrators felt that rankings of the biological sciences at Berkeley were slipping. A number of key biologists at UCB felt that the university was no longer attracting the best graduate students and faculty because the organizational structure of the biological sciences constrained the ability to develop strong faculty groups in the newly emerging areas, and because UCB's biological facilities were woefully outdated (Trow 2004). The outcome of the reorganization was the consolidation of biology at UCB into two departments in the College of Letters and Science (Molecular and Cell Biology and Integrative Biology), and the consolidation of molecular/cell research in CNR into a single department of Plant Biology (PB) (the first part of what would later become the Department of Plant and Microbial Biology). PB brought together faculty from several departments, including botany, genetics, and molecular plant biology.[3] In fact, the department central to this case study did not have a long history. The Department of Plant Biology (PB) expanded to become the Department of Plant and Microbial Biology (PMB) in 1996 (Price and Goldman 2002). This brought under one roof researchers with little previous experience working together and with rather disparate disciplinary backgrounds, ranging from botany to molecular biophysics to plant virology.

The next reorganization of CNR came in 1992. Faced with impending budget cuts, the Departments of Conservation and Resource Studies, Soils, Forest and Resource Management, Entomology, and Plant Pathology were combined into a single Department of Environmental Science, Policy, and Management (ESPM) (Stelljes 2004). This reorganization gave CNR four departments: Agricultural and Resource Economics (ARE), ESPM, Nutritional Sciences and Toxicology (NST), and PB. These four departments remain today. However, the Division of Microbial Biology was added to PB in 1996. Previously, microbial biologists were scattered across the various biology departments on campus. The 1992 reorganization was contested by a number of faculty who were moved into the new department, who felt that faculty with little in common were being grouped together in a single department. Additionally, several faculty in ESPM contend that the department has been largely dysfunctional since its inception because of its diversity and size.

It is also worth noting that a number of CNR faculty and some administrators interpreted the restructurings of CNR as an attempt to make it both more competitive and more comparable with similar colleges within and outside the UC system. As one UCB administrator argued, to be a viable department at UCB, you have to be one of the top five departments in the world. Otherwise "there is going to be serious oversight and an attempt to restructure." Consequently, departments in CNR face much competitive pressure from other colleges at UCB, most of which are highly rated at the national level, and other universities, both land grant and otherwise. PMB, for example, is often compared to MCB and other top biology programs, such as Stanford's, which tend not to be at LGUs.

While the various departments in CNR tend to have good reputations, many of them cannot easily be judged by reference to similar departments elsewhere. That is, many comparable departments are located at second- or third-tier universities, so such comparisons are not worth much, even when a UC department is judged at or near the top in comparative rankings. Furthermore, many of the faculty in departments restructured or cut altogether view the recent restructurings of CNR as privileging progressive and/or basic research over populist/applied forms of research.

CONCLUSION

Like other hot-button issues within higher education in general, the UCB-N controversy exacerbated ongoing debates within CNR and the agricultural sciences. As a result, for many of its critics, the agreement between UCB and Novartis stands for a great deal more than the issues raised by the agreement itself. For many critics, the soul and history of CNR itself are at stake. They recognize, and have more or less come to accept, that most of the work done at CNR is focused on nonpopulist issues, but they also see a long tradition of populist commitment perhaps being threatened by recent developments. Regardless of their cogency, these threats are then read into events like UCB-N. This suggests that former Harvard president Derek Bok (2003) is correct in arguing that, while the Novartis agreement is not particularly important in itself, if such agreements were to become commonplace, they would change the path of biological research in ways that might be incompatible with the goals of universities. Put another way, UCB-N and the ensuing controversy were about alternative visions of the agricultural sciences at Berkeley in particular and the future of research universities in general. In the next chapter we explore the chronology of events specific to UCB-N. In the process, we encounter actors tested by the complexity of viewpoints and faced with demands for a timely and situated judgment. Despite divisions and conflicting justifications, the pace of events and financial realities imposed tests and trials on the participants that led them to make decisions that, while not necessarily the best, are nevertheless binding.

FOUR

A Chronology of Events

EVERYTHING IS SAID to have a beginning, although reasonable people may disagree over where that beginning is. In this chapter, therefore, we present a chronology of the events relevant to the creation and implementation of the agreement between NADI and the Plant and Microbial Biology Department (PMB). For our story we think it necessary to step back a few years from the initiation of the agreement between NADI and PMB, to 1993, five years before Novartis entered this scene.[1]

In 1993 UCB Chancellor Chang-Lin Tien convened a Biotechnology Planning Board composed of ten distinguished faculty members in biological science. The intent of the board was to discuss how UCB would maintain its preeminence in biotechnology and improve technology transfer with biotechnology firms. William Hoskins, director of the Office of Technology Licensing (OTL) at UCB, was involved in this board at the request of Joseph Cerny, then vice chancellor for research. The board decided, among other things, that it wanted to have a relationship with industry patterned on the Scripps-Sandoz agreement in La Jolla, which had brought in significant industry investment. The goal for UCB was to seek $5 million per year for five or more years.[2] Although the board engaged in conversations with several companies, nothing came of it.

PLANT AND MICROBIAL BIOLOGY'S STRATEGIES

Wilhelm Gruissem, then chair of what was then known as the Department of Plant Biology, was also a member of Tien's Biotechnology Planning Board.

One administrator we interviewed suggested that Gruissem "felt that his department certainly needed the research money, so he took that concept [of significant industry investment over a considerable time period] and applied it" to developing the department. What emerged from these efforts was the formation of an International Biotechnology Advisory Board, which included fourteen representatives from industry by 1997 (Gruissem 1997). Gruissem's first tactic was to invite these company representatives to give money to support the department's graduate program in the fashion of a consortium, where each company would give approximately $10,000; but none of the companies was interested and this approach failed. Indeed, despite interest expressed by the biotechnology industry, in its three years of existence the board failed to generate any significant funding. It did, however, lay the groundwork for the subsequent agreement with Novartis by fostering departmental ties with industry.

In 1997 a research relationship between the department—reorganized into Plant and Microbial Biology—and Monsanto/Calgene was pursued within the context of the International Biotechnology Advisory Board. However, OTL discouraged PMB from creating an agreement similar to that between Monsanto and Washington University in St. Louis. There were particular terms in the latter agreement that OTL wanted to avoid, among them (1) control of patent prosecution by Monsanto, and (2) no requirement of Monsanto to make timely intellectual property (IP) licensing decisions. Carol Mimura, associate director of OTL, asserted, "Both terms would have been so far outside of University of California policies that we didn't think we could negotiate an agreement with Monsanto." When these negotiations failed, Gruissem put together a committee of four (hereafter called C4)—himself; Bob Buchanan, PMB professor; Peggy Lemaux, a Cooperative Extension specialist also in PMB; and Gordon Rausser, an agricultural economist then dean of the College of Natural Resources—and approached OTL for advice on the best way to obtain private funding for the department. Such a move had at least the tacit approval of the highest administrators within the UC Office of the President (UCOP).[3]

The ultimately successful approach C4 took to secure funding for PMB was based on Rausser's suggestion to reverse the usual relationship between funder and recipient. Under his plan, faculty would choose the source and conditions of corporate funding by pitting the companies against each other in competition for the right to collaborate with PMB. Under Rausser's plan, any university agreement with industry had to meet four basic criteria:

1. Select a single industrial partner for a research alliance that will maximize the financial, technological, and intellectual benefits for PMB, the university, and California agriculture;

2. Use traditional business models to encourage competitive bidding among candidates for partnership;
3. Insist that the strategic alliance generate large uncommitted, unrestricted funds for the department to use for research in the public interest, without oversight by the industrial partner;
4. Require that the industrial partner make significant intellectual contributions to the university in the form of access to technology and data useful for departmental research that would otherwise be prohibitive because of costs or proprietary reasons. (PMB 2002)

This reversal in the power dynamic between grantor and grantee was a major shift in the pursuit of university-industry relations and became commonly referred to—particularly by opponents of the subsequent agreement—as "auctioning off the department." A senior UCB faculty member described this shift as an effort to break what Rausser saw as the stranglehold that natural science corporations had on the direction of research.[4] A CNR professor suggested that the new approach was rooted in Rausser's argument that the "market" value of PMB could be determined only by creating a bidding process. In short, the new approach would provide a new metric for evaluating the worth of a UCB department and for comparing it with similar departments.

The "Auction" Process

A letter announcing the availability of PMB expertise was mailed to sixteen selected companies under Rausser's and Gruissem's signatures; five companies replied and were visited by C4, who in turn made presentations as to the merits of, capacities within, and opportunities available through the department. A number of guiding principles were introduced during these presentations, although not all of them were specifically incorporated into the final agreement. The principles were as follows:

- The industry partner selects participating faculty for interactions.
- In return for provision of unrestricted funds to the Department, the industry partner has free access to all uncommitted research of selected faculty.
- Funds are to be used for research by participating faculty and support of departmental and related campus programs.
- Research and publication of participating faculty remain unrestricted but are subject to review by participating industry partner.

- The industry partner obtains the right to negotiate for a license to technologies generated by participating faculty.
- Participating faculty can make individual arrangements with the industry partner to sponsor a particular program without affecting their role in the overall program.
- Participating faculty are committed to the industrial partnership program and cannot negotiate funding of a specific project with other companies unless the industrial partner relinquishes interest in the specific project.
- Industry partner may choose to have a full-time scientist located on the UCB campus. (PMB 2002; Rausser 1998)

Following the presentations, written expressions of interest were received from Monsanto, Novartis, and jointly from DuPont and Pioneer Hi-Bred.[5] Representatives from these companies then came to UCB and met with the chancellor, the vice chancellor for research, and PMB faculty. All of the participants wanted to work quickly and within a strict time frame, so PMB limited itself to a thirty-day period in which to review the proposals and select a corporate partner. Several PMB faculty members noted that each PMB faculty member had to sign a confidentiality agreement in order even to look at the competing proposals from industry. Apparently these companies promised many things verbally in an attempt to get the benefits of a relationship with PMB, but these promises were not always part of the companies' "term sheets."

Enter Novartis

In the late 1980s there was widespread enthusiasm about the potential applications of biotechnology—in pharmaceuticals, in crop and animal production, and in food processing. The nation's research universities scrambled to develop various biotechnology centers and institutes, while the large chemical and pharmaceutical companies began to invest in seed companies. There was little doubt among both supporters and opponents of these new technologies that what came to be known as the "life sciences" would never again be the same.

Leading the charge was Monsanto. Long seen as a leading chemical company, with two key patents—aspartame (NutraSweet) and glyphosate (Roundup)—providing considerable revenue, Monsanto soon captured a considerable share of the new market for genetically modified crops (Charles 2001). By contrast, Novartis was a latecomer to the field. Created in 1996 by the merger of the Swiss chemical giant Ciba-Geigy and Vienna-based Sandoz, Novartis was simultaneously among the world's largest healthcare products companies, the largest agribusiness company (including Novartis Seeds, a massive division formed by a merger between Northup-King and Ciba Seeds in 1997), and the third-largest nutritional supplements company. Moreover, it employed some 87,000 people in ten nations. It also had managed to develop and market an apparently valuable new plant product. By 1997 Novartis's *Bacillus thuringiensis* (Bt) corn (conferring insect tolerance by inserting the Bt toxin in the plant) was planted on more than 3 million acres of farmland (Kasper 2000).

Like other chemical giants such as Dow, DuPont, and Monsanto, Novartis attempted to redefine itself as a "life sciences" company. The new terminology reflected a widespread agreement that commodity chemicals were becoming less and less profitable and that research in biotechnology offered new possibilities for integrating human, animal, and plant health issues. The very name of the company—adapted from the Latin *Novae Artes* (new skills)—represented this new potential. Moreover, the company was organized into three profit centers—healthcare, agribusiness, and nutrition—even as integration was emphasized. The agribusiness division in turn was organized into three subdivisions: animal health, seeds, and crop protection. Within the seed and crop protection subdivisions, Novartis pursued goals shared widely by the industry. Specifically, like most of the biotechnology giants, Novartis would attempt to improve agricultural productivity through identification of genes conferring insect resistance and herbicide tolerance, would use bioreactors to develop new products derived from plant cells, and would identify or create valuable plant traits, such as oils with specially tailored fatty acid composition, for industrial use.

The Novartis Agribusiness Biotechnology Research Institute, Inc. (NABRI), formerly the Ciba-Geigy Agricultural Biotechnology Research Unit, had been established in 1983 in Research Triangle, North Carolina. With considerable reserves in the bank—$19 billion in 1998—and substantial revenues from its healthcare division, Novartis seemed poised to make its mark on agricultural biotechnology. Over the course of 1997 and 1998, president and CEO Daniel J. Varella recruited and hired Steven Briggs to develop a strategy for its new plant genomics research endeavor, the Novartis Agricultural Discovery Institute.

While NADI was funded at a handsome $600 million over ten years, it was clearly dwarfed by the $8 billion spent by Monsanto between 1996 and 1998 (Kasper 2000). NADI was a wholly owned entity of the Novartis Research Foundation, itself controlled by the Novartis Corporation. But since Novartis was a latecomer to the plant biotechnology arena, it decided to take a different approach to the field from that taken by Monsanto, DuPont, and others. Specifically, NADI would become the Bell Labs of plant biotechnology. It would engage in fundamental research considerably removed from the product development common at other Novartis labs. The coincidence of the establishment of NADI and the opportunity to collaborate with PMB proved fortuitous for both institutions.

A study by the Scottish Universities Policy Research and Advice Network at the University of Edinburgh, based on executive interviews conducted in 1999, summarized the distinctions between the two research institutions:

> In strategic planning for new product development the next ten years is irrelevant for R&D decision making. Any decisions taken for research today are for after 2010. In Novartis Agribusiness 'the present day' covers the period up till 2005, mainly the operating companies engaged in product development. 'Tomorrow', covers time lines approximately 2005–2010, roughly, and involves research conducted at Novartis Agricultural [Agribusiness] Biotech Research Institute (NABRI) in North Carolina. 'The day after tomorrow', covers 2010 and beyond and involves mainly the Novartis Agricultural Discovery Institute (NADI) in California. But these dates are approximate and the technology can surprise a company positively or negatively. (Tait and Chataway 2000, 22)

In addition, unlike its competitors, Novartis would attempt to develop partners outside the company that might help it to advance its goals. As one observer put it, "Under this model, Novartis would neither attempt to buy critical technologies from outside, nor build them in-house, but rather acquire them through a web of relationships with external companies and academics. Briggs would design his organization to appeal to top researchers, many of whom might not feel at home in a large corporate bureaucracy. His organization would acquire some of its own researchers, but also function as the nexus of a larger maze of interconnected, but independent, scientists" (Kasper 2000, 7). This approach was far cheaper than that of Monsanto, which in 1998 had bought DeKalb Seeds for $5.4 billion. Briggs determined that the best location for NADI was San Diego, where likely collaboration between Novartis and Scripps Research Institute could be undertaken. Furthermore, the excellent weather in the San Diego area would help to attract a top-flight research team (Kasper 2000).

Indeed, one difference in the proposals received by PMB is reported to have been that Novartis was "the most accommodating of the university spirit," including academic freedom and the principles set out by PMB.[6] This may have been rooted in Novartis's relatively weak position in plant biotechnology in 1998 and its desire to establish university-industry synergies to foster the development of NADI.

Although the relationship with Scripps never materialized, PMB's solicitation arrived at precisely the right moment. Steven Briggs and Wilhelm Gruissem had known each other for some time, which undoubtedly also helped.
Other connections contributed to the selection of NADI as well. Steven Briggs
had been working at Pioneer when PMB began its search for an industrial
partner. Briggs was not the only person to move from Pioneer to NADI.
Indeed, when the four representatives of PMB came to present their plan to
NADI, they found many people they had seen a few months earlier at their
Pioneer presentation.

Given the noticeably different character of the Novartis proposal—
combined with the fact that Novartis offered more money than did Monsanto or DuPont/Pioneer—the faculty decided to pursue negotiations with the
company. Put more precisely, and as noted above, PMB did not enter into an
agreement with the Novartis Corporation per se but rather with NADI, which
was formally funded by the legally distinct Novartis Research Foundation.[7]

Negotiations

Once PMB decided on Novartis in May 1998, they wrote a letter of intent to
the company specifying a thirty-day time frame within which the agreement
would be written. PMB conferred with OTL during these discussions in order
to clarify issues with respect to university policies and procedures, but prior
to this they did not work closely with each other. Indeed, PMB had intended
to hire an outside attorney to write the agreement, but Vice Chancellor Cerny
argued that the agreement should be drafted by OTL, and by Hoskins in particular. The concern was that an outside attorney would not be familiar with
the aims of UCB; the agreement was expected to have characteristics not typical of university-industry agreements; and, in addition, there was a rigid time
frame of thirty days.

At this point a meeting took place between Joyce Freedman, director of
the Sponsored Projects Office (SPO), Cerny, and Hoskins, at which Cerny
made it clear that he wanted Hoskins to write the agreement. This meeting
was necessary because the usual route for university-industry contracts began
with SPO. OTL was to be involved only later, should IP issues need to be
resolved. Hoskins was worried about this inversion and the perception that
OTL was stepping on SPO's territory, and he wanted Cerny to say publicly that
he, Cerny, wanted it done this way, with OTL's involvement from the start.

There was insufficient time to write a full comprehensive draft agreement
because PMB had promised one to Novartis within thirty days. So Hoskins
started with a model agreement that UC, San Francisco, had signed with

Daiichi Pharmaceutical Corporation, which was smaller in both scope and dollars. Hoskins also looked at the UC, Irvine–Hitachi contract, which included the construction of a building. A team of people drafted the agreement between UCB and Novartis: three PMB scientists, two senior representatives from NADI, two senior administrators from OTL, and two attorneys from Novartis. Perhaps remarkably, they managed to stay within the thirty-day time frame and had the first draft of the agreement written by June 1, 1998. Once drafted, it was passed on for review to the UC Office of the President, the UC general counsel's office, and Novartis's corporate headquarters in Basel, Switzerland.

Negotiations continued on the structure and content of UCB-N as each party received feedback on the draft agreement. In September 1998 Hoskins and Carol Mimura (associate director of OTL) gave a number of presentations on the draft agreement to the assembled PMB faculty. Hoskins described these presentations as detailed and frequent: "Carol and I would go over there at about three o'clock in the afternoon, and we stayed as long as it was necessary, and we went over every article of the agreement." He recalled that there were approximately four meetings, each lasting two or three hours, with thirty or forty faculty members, and that "none of them had ever seen an agreement, I don't think, of that magnitude or that complexity."

During the negotiation of the formal contract there was talk of a second $25 million from Novartis for a research facility to be built on campus, along with the $25 million research grant. This facility was proposed to enable closer collaboration between the PMB and NADI scientists. Although the interface facility was discussed from the very beginning, Hoskins said, "That was nice in principle, but I could see after about the first month that it wasn't going to happen in my lifetime. So I set that aside and I said, 'we need to focus on getting the research agreement done because that's important to the faculty.'" Indeed, Appendix E of UCB-N, "Guidelines for working in NADII-UCB facility," states that "the development of these Guidelines is deferred until the NADII-UCB facility is identified" (NADI and UC 1998, 44).

Identifying a suitable site for the interface facility proved to be an insurmountable hurdle. The ideal site had to be close enough to campus that people could walk to it, and it had to be built or renovated quickly so that the participants in UCB-N could use it for the bulk of the five years covered by the agreement. Although several sites were considered, the process simply took too long. One year after UCB-N was signed, NADI withdrew the money and the idea was quietly abandoned.

Another part of the negotiations concerned whether USDA employees at the Plant Gene Expression Center (PGEC) in Albany, California, would be involved in UCB-N through their adjunct status with PMB. Athanasios

Theologis, a PGEC scientist, said that Novartis sought to include the PGEC employees because of their excellence in research. However, the USDA hierarchy was unhappy with their employees participating directly in UCB-N. Bob Buchanan recalled, "Some high-ranking people from the USDA came out and visited when this [the agreement] was beginning to gel, and they just couldn't commit to one corporation, right, because of the government position." Interestingly, according to Peter Quail, research director at PGEC, UCB-N was the only grant gained through adjunct status at UCB to which the USDA objected. In the end, PMB returned $1,056,000 of the $25 million from NADI because the PGEC scientists could not be involved in the agreement. In turn, each PGEC principal investigator (PI) who wanted to be involved then negotiated an individual Cooperative Research and Development Agreement (CRADA) with NADI to have access to funds, equipment, and databases.

In sum, from the perspective of PMB faculty, the agreement with Novartis was expected to contribute to the common good. It would support the three central principles quoted above. It would provide a high level of autonomy to faculty to pursue their research interests. It would promote (or at least not stifle) creativity, as well, by minimizing the paperwork and reducing the time needed to obtain grants of significant size. And, by involving nearly the entire department, it would promote a sort of diversity—diversity with respect to research approaches and standpoints of PMB faculty.

Furthermore, the agreement combined the three forms of distribution discussed earlier in a novel way. Funds would be obtained through free exchange (via the "auction" process). The need for research funds would be met, and those who deserved the funds would receive them on the basis of peer review of proposals. But the concerns raised at PGEC presaged broader and more challenging concerns on the part of UCB faculty and students as well as external stakeholders.

EMERGING OPPOSITION

The main part of the negotiations between PMB and Novartis occurred during the summer months, when few faculty members or students were on campus. With the arrival of the fall semester, however, rumors began to circulate about an alliance CNR was forming with Novartis. Several of the first-year students in the Department of Environmental Science, Policy, and Management (ESPM) in CNR were particularly incensed upon hearing of the potential alliance. They formed an organization called Students for Responsible Research (SRR) and sought further information. Many CNR faculty members, along with individuals across campus, were also concerned about the process through which UCB-N was being created, the repercussions of this agreement as it was put into practice, and the implications for the future of the university.

On October 2, 1998, Rausser held a CNR "town hall" meeting. Multiple attendees (faculty, students, and staff) have described that meeting as a forum held to dispense information about a fait accompli rather than as an arena in which to solicit advice. Several of the faculty members who attended this meeting were struck by how few people spoke out, even though there were widespread indications that many faculty members privately had strong reservations about the agreement.

One of the main points of contention that surfaced during this meeting, and that informed much of the subsequent controversy concerning UCB-N, was the precise academic unit with which NADI was entering an agreement. Up to this point it was a PMB agreement, but there had been a deliberate decision to formally extend it to CNR as a whole. In retrospect, almost everyone in PMB thinks that this was a serious mistake. The idea was Rausser's, and he argued for a formal extension to the college for three reasons: (1) he saw his own involvement in negotiations with companies raising the credibility of, and potential resources offered by, PMB; (2) he anticipated that any negative criticism could be directed at CNR rather than at PMB, and CNR had more resources for handling the media and any questions; and (3) he knew that CNR desperately needed unencumbered funds—particularly for infrastructural and renovation work—that would come from its share of the indirect costs. By this time, however, the agreement was in the final stages of negotiation, and it was too late to include the participation of other CNR faculty members.

At the town hall meeting, Rausser outlined the draft agreement and, according to David Quist, ESPM student and member of SRR, said there was no secrecy surrounding the agreement and that the draft was available in the dean's office for anyone who wanted to review it. Four days later, on October 6, Quist visited Rausser's office to look over the draft. All he found were Rausser's outlined notes on a presentation he had given to a commodity group on potential strategic alliances, a presentation that contained no contractual language on the agreement with Novartis. Quist attempted to take notes on what he saw, but his notes were confiscated by the dean's office staff. In an interview with our team, Quist argued, "At that point, obviously, that kind of signaled that the process wasn't as transparent as he [Rausser] was trying to making it seem it was."

SRR continued to search for reliable information about UCB-N through both formal and informal channels. Largely unsuccessful in this effort, the group turned its attention to a media campaign that would seek broader public involvement and commentary. On October 14 SRR presented the UC

Regents with a petition signed by four hundred people, predominantly graduate students in ESPM, along with some undergraduates and staff members from across the campus and a handful of people unaffiliated with the university (Students for Responsible Research 1998a). The petition asked for a delay in formally signing UCB-N until various concerns had been addressed. It was effectively ignored.

After SRR had approached several newspapers with their reservations about UCB-N, the organization met with Rausser on October 16. At this meeting Rausser made a public show of returning Quist's notes to him and tried to assure SRR that the agreement with Novartis was fine. In the students' view, however, and given that no agreement or agreement summary was ever produced, their questions remained unanswered.

At the same time that SRR was developing its opposition, the executive committee of CNR (ExCom) also became actively involved in the controversy. ExCom had elected a new chair, Ignacio Chapela, an untenured assistant professor in ESPM, just one week before getting notice that UCB-N was in the works. In 1997 there had been some theoretical discussions in ExCom regarding university-industry contracts in CNR, but this was the first time they had dealt with any specifics. Rausser brought the tentative agreement to Chapela in the hours before the ExCom meeting and asked that it be presented to the committee for their approval. Chapela was reluctant to seek approval for an unscheduled, unread, and undiscussed agenda item. In any event, ExCom did not approve the document, as several committee members expressed reservations about numerous aspects of UCB-N after quickly looking it over. Andrew Gutierrez, professor of ecosystem science, said, "Since we [ExCom] are advisory, I think that it was important for him [Rausser] to have some stamp of approval of the faculty, and we, being representatives of the faculty, could have easily legitimized it. And in one fell swoop it would have gone through. And then maybe a year or so the later the faculty would start raising hell themselves. It didn't happen that way. We questioned it right away."

ExCom subsequently distributed a survey to CNR faculty regarding the pending agreement (CNR ExCom 1999). This survey was based on concerns and questions that faculty had raised in informal meetings and conversations, to which they had received partial or unsatisfactory answers from the dean's office. While the survey had a response rate of 59 percent, it was discarded owing to a potential failure to maintain anonymity. (The survey failed to follow the common procedure of using two envelopes, one inside the other, only the outer of which would have a name attached.) This meant that there was no representative assessment of CNR faculty opinion regarding the agreement prior to its inception.

INVOLVEMENT OF THE ACADEMIC SENATE

There are at least three different accounts of how the Berkeley Academic Senate was informed about, and came to be involved in discussions concerning, UCB-N. The documentation we were able to gather indicates that Rausser contacted Robert Brentano, then chair of the Divisional Council of the Academic Senate (DIVCO), in mid-August 1998 to request that the Academic Senate consider the potential difficulties that might arise (Brentano 1998a). This request marks a point of departure from the usual protocols, as the Senate had never before been consulted about a forthcoming contract with industry, and, according to the policies of UCB, Rausser was under no formal obligation to inform the Senate about a potential alliance between the university and a private company.

As it was, Brentano was not familiar with university-industry relations and delegated to Todd LaPorte, then chair of the Academic Senate's Committee on Research (COR), the task of leading a working group composed of Robert Spear, then vice chair of DIVCO, and four other chairs of the relevant Senate committees: David Littlejohn (Academic Freedom), David Hollinger (Budget and Interdepartmental Relations), Richard Fateman (Academic Planning and Resource Allocation), and John Lindow (Graduate Council). From September 1998 on, the Berkeley Academic Senate closely followed the UCB-N negotiations and the growing concerns of people across campus.

The COR working group gathered what information was available at that time from C4. This information was given in turn to other Academic Senate committees and to CNR's ExCom. It was also discussed at two DIVCO meetings. By September 30 the working group had compiled thirty-seven questions, some with many subquestions, regarding UCB-N. Further, the group drafted a proposal to treat UCB-N as an experimental form of university-industry relations that required further study. Brentano, as chair of the Academic Senate, sent this document to Carol Christ on October 6 (Brentano 1998b) and received a response on October 16 (Christ 1998a). The Committee on Academic Freedom expressed its disappointment in the administration's response in the strongest possible terms: "[W]e were frankly disappointed in Vice Chancellor Christ's reply ... we were surprised to see the Vice Chancellor dismiss all of our concerns out of hand. ... The most generous characterization we could give to many of these answers was 'evasive': they simply eluded or ignored the point of the question; or juggled statistics, assured us all would be well, or informed us we didn't know what we were talking about" (Littlejohn 1998).

On November 18, based on the feedback from DIVCO and various Senate committees, Brentano (1998c) wrote Christ that "the Senate cannot fully endorse the Novartis agreement at this time as there are core issues which have not been, and perhaps cannot be, adequately addressed." The two main sticking points were the lack of a faculty member without ties to CNR or Novartis on the UCB-N Advisory Committee,[8] and the need to treat UCB-N as an experiment and conduct an "on-going assessment of the institutional impact of the agreement." Less than one week later, Christ (1998b) agreed to the Senate's demand for an assessment and appointed the Center for Studies in Higher Education (CSHE) to direct it.

Signing of the Agreement

At the end of October 1998 Hoskins left for a two-week trip to Hungary, and Carol Mimura handled the final rounds of UCB-N negotiations. After UCB-N was signed, OTL's role was effectively complete, except to the extent that discoveries and potential patents arose, issues that were subsequently assigned to Mimura. Given the draft character of the original agreement, OTL expected changes in the contract to be necessary. Hoskins said, "If we had stopped and waited to get the approval of everybody that we needed to get approval for, in writing, probably the agreement wouldn't be done today." As it turned out, though, there have only been three minor amendments to UCB-N.[9]

The agreement was more or less ready to sign in early November, but, perhaps in response to the emerging controversy, UCOP sent new attorneys to review the agreement at this point. These attorneys were not versed in IP law and also had to be brought up to speed on the history of the negotiations, points of law, and the background of the agreement. This pushed back the formal signing a few weeks.

At 11:00 A.M. on Monday, November 23, 1998, the institutional representatives of UCB and Novartis held a joint press conference in Koshland Hall, the home of PMB, at which the agreement was officially signed (UCB Public Information Office 1998). Attending for UCB were Robert M. Berdahl, UCB chancellor, Joseph Cerny, UCB vice chancellor for research, and Gordon Rausser, dean of CNR. Douglas G. Watson, president and CEO of the Novartis Corporation, and Steven P. Briggs, president of NADI, represented the company. Various people from print and radio media were present to cover this event. They were able to report on the opposition's response as well, which came in the form of a pie thrown at Rausser and Cerny that hit only Cerny. Immediately following the press conference SRR held its own press conference, at which the student group denounced UCB-N as an "improper relationship between a public institution and a private company" (Students for Responsible Research 1998b).

Continued Questioning

The story did not end with the signing of the agreement. Indeed, in many ways the official establishment of UCB-N only marked the end of the first chapter of the story. With the start of a new semester in January 1999, ExCom sent out a second survey in an effort to get CNR faculty opinion on UCB-N. The format for this survey was slightly modified from the earlier one (some questions were reworded because the agreement had already gone into effect), and the distribution method was revised to prevent the kind of criticism that marred the first attempt. The response rate was an acceptable 68 percent, in comparison to 59 percent for the first survey (CNR ExCom 1999). Among other things, the results from the second survey indicated the extent of division among the departments that made up CNR. While the majority (72 percent) of CNR faculty who responded agreed that biotechnology was an appropriate area of research, there was disagreement over the appropriateness of various funding structures within which this research should be performed. The responses from PMB faculty were notably different from those of other CNR faculty. PMB faculty overwhelmingly supported all forms of university-industry funding arrangements. Moreover, PMB was the only department in which the majority of the faculty members considered the use of university facilities by industry scientists appropriate. Not surprisingly, PMB was most supportive of the agreement and expected it to have mostly positive consequences; in contrast, ESPM faculty (specifically the divisions of ecosystem science, and resource institutions, policy, and management) were most critical of the agreement and predicted mainly negative effects.

Eighteen months later, on May 15, 2000, the California State Legislature also questioned the agreement. The Senate Committee on Natural Resources and Wildlife and the Senate Select Committee on Higher Education held a joint hearing chaired by Senator Tom Hayden entitled "Impacts of Genetic Engineering on California's Environment: Examining the Role of Research at Public Universities (Novartis/UC Berkeley agreement)" (2000). The involvement of the state senate in a review of a particular university-industry contract was perhaps unprecedented and at the very least highly unusual. Indeed, such an involvement raised concern in the UCB Academic Senate. Ronald Amundson (2002, 3), chair of the Committee on Academic Freedom, said "The choice of research directions and the way it is developed *must* be initiated and controlled by the faculty of the University, not by outside institutions.

… At the very least, direct legislative investigations of individual UC departments and programs *must be vigorously opposed by the Academic Senate.*"

In sum, the many disparate voices of discord challenged the actions of those who supported UCB-N. They implied a different view of the common good, different (indeed, divergent) views of the three central principles of the university, and different views of the distributive aspects of UCB-N. They questioned whether the agreement permitted sufficient autonomy to those involved. They asked whether it might stifle the creativity of PMB faculty in subtle and perhaps invisible ways. And, they suggested that the involvement of CNR as a party to the agreement might actually reduce intellectual diversity by shifting resources toward molecular biology and away from other fields of endeavor.

COMMISSIONING THE EXTERNAL STUDY

While those faculty who signed on to UCB-N got down to the business of writing proposals and generally implementing the agreement, a subcommittee of the Academic Senate's COR debated throughout the spring semester how the independent study of UCB-N would be structured. This subcommittee took over much of the work that COR and the working group had done on UCB-N and comprised many of the same people. At the end of May 1999, Jean Lave, then vice chair of COR and professor of social and cultural studies, offered to write the request for proposals to which research teams would respond. After negotiations with Christ, "it was informally agreed that due a) to the timing of the first annual disbursement of $5 million of Novartis funds in Spring 1999, and b) to the imperative of developing at least some credible baseline data on the existing situation, the EVC [executive vice chancellor's] Office would make available resources for one month's summer salary and research assistance to enable Professor Jean Lave to assemble an initial data set on conditions in the College" (LaPorte 1999).

With this support Lave hired Gwen Ottinger, a doctoral candidate in the Energy and Resources Group, and together they formally interviewed nineteen people over the course of five months, held informal discussions with many others, and collected various university documents (Lave 1999). At the beginning of November Lave drafted a request for proposals based on this information that sought "a proposal for a single, interdisciplinary team research project, whose PIs are from outside the Berkeley community" (Lave 1999, 1).

At this point EVCP Christ asked Anne MacLachlan, a senior researcher at CSHE, if she would interview PMB graduate students to get some baseline data on how they saw and were affected by UCB-N, and to name how much money she needed. With six months' worth of funding, and Lave's one month of summer salary paying for the assistance of Mary Crabb, a graduate student in social and cultural studies, MacLachlan and Crabb conducted thirty-five interviews with PMB graduate students during the first half of 2000 (MacLachlan 2000).

After giving the chancellor's office a draft request for proposals, and while MacLachlan conducted her study, Lave went to Copenhagen for six months to conduct research for an unrelated project. When she returned she found that very little had happened to move the external study of UCB-N closer to fruition. In June 2000 Lave contacted John Cummins, associate chancellor and chief of staff to Berdahl, and explained the probability of negative consequences for the university if their public promises of an independent study were not kept. Indeed, according to Spear, the Academic Senate had negotiated with the administration for an external review because many of the campus concerns regarding UCB-N were "what if" questions, and simply could not be answered a priori. The promise of an independent study had contained the faculty controversy; as one faculty member remarked, "if there had been no agreement like this [on an external review], then there would have been hell to pay. There was enough faculty concern that I think there would have been a lot of trouble had it not been for the administration's agreement to actually investigate the impact of this thing." But those working hardest to get the external study going thought they were being stonewalled by the administrators most closely involved in negotiating UCB-N, who, they thought, did not appreciate the importance of the study. It took two and a half years to get an adequate budget and enough commitment for the review even to get off the ground.

On July 1, 2000, as negotiations continued, Richard Malkin succeeded Gordon Rausser as acting dean of CNR and Paul Gray replaced Carol Christ as EVCP. It appears that by this point enough pressure had been brought to bear on the administration to keep its promise to treat UCB-N as an experiment that the external study was finally ready to commence. At the beginning of November 2000, Gray held a meeting to establish an oversight committee to recruit and select the PI for the external study (Upshaw 2000). Chancellor Emeritus Ira Michael Heyman, then acting director for CSHE, agreed that the CSHE would be the source of administrative support and guidance for both the oversight committee and the external research team. This committee was chaired by Professor Emeritus Michael Teitz, a former chair of the Academic Senate and director of research and senior fellow at the Public Policy Institute

TABLE 4.1   Chronology of Events

1993	UCB Chancellor Chang-Lin Tien established the Biotechnology Planning Board.
	PMB established the International Biotechnology Advisory Board.
1997	Representatives from PMB consulted with the dean of CNR, Gordon Rausser, regarding how best to procure industry funding for PMB.
	The "committee of four" (C4) was established in PMB.
December 1997– January 1998	PMB contacted nine companies, indicating that a large group of PMB faculty was interested in seeking industry funding from a single company. Six companies responded.
February–April 1998	C4 made presentations to the six companies that responded to its proposal.
April 30, 1998	PMB received four proposals for a joint partnership.
May 1998	PMB chose Novartis for its industrial partner.
	Novartis approved the proposal to construct NADI.
June 1, 1998	The first draft of the agreement was completed.
August 1998	The Academic Senate was contacted regarding the pending agreement between PMB and Novartis.
September 1998	William Hoskins and Carol Mimura presented a draft of the agreement to PMB.
October 1998	ExCom of CNR conducted the first survey of CNR faculty members.
October 2, 1998	Rausser held a "town hall" meeting to present the agreement to the UCB community.
October 6, 1998	Robert Brentano, chair of the Academic Senate, sent a document outlining the Senate's concerns regarding the pending agreement to Carol Christ, EVCP.
October 14, 1998	SRR presented the UC Regents with a petition of 400 signatures asking that the signing of the agreement be delayed until various concerns were resolved.
October 16, 1998	Carol Christ replied to the Academic Senate regarding their concerns about the pending agreement between PMB and Novartis.
November 18, 1998	The Senate stated that it could not fully endorse the agreement, that it wanted the agreement to be treated as an experiment, and that an ongoing assessment of the agreement's impacts should be undertaken.
November 22, 1998	Carol Christ agreed to the Senate's stipulations and appointed CSHE to direct the assessment of the agreement.
November 23, 1998	The agreement was formally signed.
January 1999	ExCom conducted its second survey on faculty opinions of UCB-N.
	NADI paid the first installment of funds for UCB-N.
November 1999	Jean Lave created the first draft request for proposals for the external study.

TABLE 4.1   Chronology of Events *(continued)*

February–June 2000	Anne MacLachlan interviewed PMB graduate students on their opinions of UCB-N.
May 15, 2000	The California state legislature held a hearing to review the agreement.
July 1, 2000	Richard Malkin succeeded Gordon Rausser as acting dean of CNR.
	Paul Gray replaced Carol Christ as EVCP.
	Brian Staskawicz replaced Wilhelm Gruissem as UCB-N PI.
November 2000	The Novartis Oversight Committee, Michael Teitz as chair, was established to oversee the selection of the PI to do the external study of UCB-N.
	Novartis spunoff its agribusiness division and combined with the agribusiness division of Astra Zeneca to create Syngenta.
January 2001	Syngenta bought NADI, renamed it TMRI, and took over the agreement with PMB.
	Beth Burnside replaced Joseph Cerny as vice chancellor for research.
August 2001	Letter of invitation sent to prospective PIs by Michael Teitz for the Novartis Study Committee.
September 2001	Proposals submitted to Novartis Study Committee.
October 2002	Release of the internal review conducted by the office of the vice chancellor for research.
November 23, 2003	UCB-N expired.

of California. The oversight committee represented a range of conflicting viewpoints regarding UCB-N.

In order to carry out its charge the oversight committee had to decide what leadership qualities it wanted, how to inform potential leaders that the study was ready to be undertaken, how to select candidates, and, finally, how to oversee the implementation of the external study. Discussions were lengthy, but at a meeting on September 10, 2001, an external review team from Michigan State University (MSU) was selected to pursue the study. SPO and CSHE argued over the available financial resources until December 20, 2001, when the contract was finally sent to MSU's grants and contract office.

The oversight committee dissolved once its mission was fulfilled with the selection of MSU to conduct the external study of UCB-N. At this point another committee, initially also called the oversight committee, was formed by EVCP Gray to act as a liaison between UCB and the MSU team. This second committee was chaired by Robert Spear and consisted of six people, most of whom had long been interested in the issues raised by UCB-N. (For a schematic chronology of the long negotiation process, see Table 4.1.)

RECONFIGURATION OF NOVARTIS
AGRICULTURAL DISCOVERY INSTITUTE

While all of these meetings and negotiations were taking place on campus, the industrial partner in the contract morphed into a new entity. By 2000 the market for agricultural biotechnology had substantially contracted. A combination of longer lead times for product development, public opposition to genetically modified foods, lower returns on investments when compared to the highly lucrative pharmaceutical industry (Deutsche Bank 1999), and the collapse of the "life science" strategy (Assouline et al. 2002) led to the divestment of agricultural biotechnology companies. Pharmacia, the successor to the Upjohn Corporation, sold off Monsanto. Novartis merged its interests with Zeneca Agrochemicals to form Syngenta AG. NADI was renamed the Torrey Mesa Research Institute and initially remained a key part of Syngenta's corporate strategy.

But, much as the breakup of AT&T made basic research at Bell Labs too costly to maintain, so the separation of Syngenta from Novartis meant that the company's commitment to basic plant science research would be abandoned.

According to filings with the U.S. Securities and Exchange Commission (SEC), Novartis's agribusiness activities made up 22 percent of its business activities before the separation (year ending December 31, 1999). In the same year Novartis completed major acquisitions in the areas of animal health, pharmaceuticals, and specialty lenses. Syngenta AG, which is headquartered in Basel, Switzerland, has its largest sales from selective herbicides designed to control weeds in corn, cotton, rice, sorghum, soybeans, sugar cane, and wheat. Fungicides and insecticides are the next-largest product groups. Syngenta's rather lackluster financial performance during the term of the agreement is summarized in Table 4.2.

As noted above, while NADI was affiliated with Novartis it was funded by the separate Novartis Research Foundation, but with the spin-off of Novartis's interests in agricultural biotechnology NADI was no longer of use to the parent company. Syngenta formally purchased NADI and changed its name to the Syngenta Agricultural Discovery Institute, Inc., before changing it again to the Torrey Mesa Research Institute (TMRI). Section 19 of the original UCB-N contract allowed for the agreement to be continued by any industrial successor. Furthermore, Syngenta's mission for TMRI was similar to that of NADI: long-term (ten or more years) research for future product introductions.

However, Syngenta announced the closing of TMRI in December 2002. A portion of TMRI's research activities continued in California through its sale of technology and IP to the biotech startup Diversa Corporation in San Diego,

TABLE 4.2   Syngenta Financial Highlights, 1999–2004

Fiscal year	Sales ($ billions)	Net profit ($ millions)
1999	4.678	135
2000	4.876	564
2001	6.323	34
2002	6.163	220
2003	6.525	340
2004	7.269	762

Source: Syngenta annual reports.

in which it held 14 percent equity (Syngenta 2002), for $39 million (pretax) (Syngenta 2004). According to SEC filings by Diversa, on February 20, 2003, Syngenta committed to provide at least $118 million of research contracts to Diversa over seven years. The former CEO of NADI and TMRI, Steven Briggs, joined Diversa as vice president for research and development as well (Vogel 2002). In addition to Syngenta, at the time of the agreement Diversa had alliances with Dow Chemical Company, DuPont Bio-Based Materials, Givaudan Flavors Corporation, GlaxoSmithKline, and Invitogen Corporation.

Syngenta consolidated its remaining interest in agricultural biotechnology into its main U.S. operations in North Carolina. Some eighty positions in California were eliminated and thirty were moved to North Carolina. Seventy-six employees were to move to Diversa. As Dr. David Lawrence, head of Syngenta Research and Technology put it, "The agreement [between Syngenta and Diversa] enables us to broaden our biotechnology capacity and bring innovative products to market more quickly" (Syngenta 2002, 1). In short, Syngenta retreated from the long-term research initially proposed for NADI. Nevertheless, it did honor the remaining portion of its $25 million commitment to PMB. But the closure of TMRI, the decline in the fortunes of the company, and the failure of the agreement to generate any patent licenses made renewal of the grant impossible.

UNIVERSITY OF CALIFORNIA, BERKELEY
INTERNAL REVIEW

At the beginning of October 2002, the office of the vice chancellor for research disseminated the results of an internal administrative review of the agreement (Price and Goldman 2002). During the exchange between DIVCO and EVCP prior to the signing of UCB-N, Christ (1998b; 1999, 13) agreed to such a review, to be performed at the halfway point in the five-year agreement so that any necessary adjustments discovered by the review could be incorporated for the second half of the contract. The review's authors did not recommend any adjustments in the language of the contract or in its implementation, however, concluding that "virtually none of the anticipated adverse institutional consequences has been in evidence" (Price and Goldman 2002, 39).

The review was circulated among members of the Academic Senate and four of its committees (Academic Freedom, Research, Academic Planning and Resource Allocation, and the Graduate Council). The Graduate Council chose to respond formally at the February 10, 2003, meeting of DIVCO. While its members agreed that the review was informative, all of the committees were disappointed with its narrow focus and its unwillingness to generate broader conclusions about future university-industry relations (DIVCO 2003).

IMPLEMENTATION OF THE AGREEMENT

While COR and the Academic Senate strove to bring about an external study of UCB-N, those intimately involved in the daily implementation of the agreement got down to work. The first proposals from the faculty participants to the UCB-N Committee on Research (COR)[10] were submitted at the end of 1998, and Novartis's first installment was paid in January 1999. After that, proposals were submitted annually to the COR in the first half of October to give the COR sufficient time to review the proposals before the next funding year began. All of the proposals were funded at the amounts requested, and although the level of funding for some faculty members changed from year to year, the average annual award was $120,500, with a range of $60,000 to $200,000. Twenty-five PMB faculty members received money from UCB-N, but two left PMB during the agreement and therefore were not funded for all five years. PMB hired another faculty member late in the agreement and he received UCB-N funding for only the last year.[11]

Wilhelm Gruissem, the prime mover behind the agreement and its PI, was one of the faculty members who left. On July 1, 2000, he took a position as professor of plant biotechnology at the Swiss Federal Institute of Technology in Zurich. Upon Gruissem's resignation from PMB and as PI of UCB-N, Brian Staskawicz was appointed PI of the agreement.

During the period from November 23, 1998, to November 23, 2003, PMB made fifty-one disclosures to OTL, twelve of which resulted from research funded solely by NADI. In total, twenty of the PMB disclosures were patented, ten of them resulting at least in part from funding provided by UCB-N.[12] Six of the patented disclosures were pursued by NADI, but no options to negotiate an exclusive license were executed.

In addition to the unstructured collaborations that occurred between PMB and NADI scientists via e-mail, telephone, and visits to the NADI site in La Jolla, there were also formal annual retreats for the participants of UCB-N. The first of these took place in January 1999 and served as an opportunity for everyone involved to meet one another and learn of one another's academic interests. There was a second meeting in October 1999 that took the form of a workshop in which affinity groups of PMB faculty and NADI personnel formed around particular topics of mutual interest. For the remainder of the agreement, annual retreats occurred in October and provided updates on the research sponsored by NADI. The last retreat was held on October 23, 2003, exactly one month before the agreement ended.

AFTERMATH

Although not directly connected to UCB-N, two other sets of events took place that many interpreted, rightly or wrongly, as consequences of the agreement. The first concerned the handling of Professor Ignacio Chapela's tenure case. The second concerned the subcontract between Professor Tyrone Hayes and a consulting firm with a contract from Syngenta. In both instances the existence of UCB-N certainly influenced the process and perhaps the outcomes. Let us consider each in turn.

Ignacio Chapela's Tenure Case

As noted above, Ignacio Chapela, an untenured faculty member in ESPM, was appointed chair of CNR's ExCom in 1997, just before UCB-N became a public issue. Chapela himself was a highly vocal public critic of the agreement, both in his role on ExCom and as a concerned faculty member. As the chair of ExCom, Chapela also headed the faculty survey on UCB-N, which itself became quite controversial.[13]

In the November 29, 2001, issue of *Nature,* David Quist, a graduate student who was also a critic of UCB-N, and Chapela published an article alleging that native maize landraces in Oaxaca, Mexico, contained introgressed transgenic DNA constructs (Quist and Chapela 2001). Furthermore, they argued, the transgenes were unstable. To some observers, this implied the destruction of peasant landraces that might be used in maize improvement for future generations. Their paper generated far more criticism than the vast majority of scientific papers ever do.

As Latour (1987) notes, the fate of most scientific articles that are deemed weak is to be ignored. Such was not the case for Quist and Chapela's piece.

This suggests that far more was at stake than biological theory and scientific integrity. Debate exploded in the scientific world, and especially in the biotechnology sector, over Quist and Chapela's findings. Their paper probably would have attracted little attention outside the field had it not been for a heated debate that began on the AgBioWorld website (http://www.agbioworld.org). While the British newspaper *The Guardian* provided evidence asserting that the debate was initiated by several fictitious scientists traced to a public relations firm hired by Monsanto (Monbiot 2002), the main protagonists in the debate were several members of PMB. The debate ultimately resulted in something resembling a retraction by *Nature*, which published an editorial note reading,

> [W]e received several criticisms of the paper, to which we obtained responses from the authors and consulted referees over the exchanges. ... In light of these discussions and the diverse advice received, *Nature* has concluded that the evidence available is not sufficient to justify the publication of the original paper. As the authors nevertheless wish to stand by the available evidence for their conclusions, we feel it best simply to make these circumstances clear, to publish the criticisms, the authors' response and new data, and to allow our readers to judge for themselves. (Campbell 2002, 600)[14]

In September 2001 the tenure review process for Chapela by his colleagues in ESPM began. As is typically the case in university tenure decisions, the precise details of the debate over tenure were not made public.[15] However, a large—some might say an extraordinarily large—number of letters were solicited from external reviewers of Chapela's work. As far as can be discerned, they were overwhelmingly positive about his work. Moreover, in the spring of 2002 the ESPM faculty voted 32-1, with three abstentions, in favor of tenure. The case was then forwarded by the dean to the campus ad hoc tenure committee for further review.

In an unusual move, the chair of Chapela's ad hoc tenure committee, Steve Beissinger, asked that those involved in the review have nothing to do with UCB-N, ostensibly as a result of the controversy over UCB-N and the ensuing dispute over the *Nature* article. The ad hoc committee recommended tenure unanimously on October 3, 2002.

At this point, according to some observers, the vice provost asked the ad hoc committee chair to reevaluate the case with additional external letters. Soon afterward, the chair of that committee resigned, renouncing the report of his own committee and citing a lack of expertise on the kind of research that Chapela conducted. A member of the ad hoc committee, Wayne Getz, called Chapela's tenure review disgraceful (Walsh 2004).

The dossier was then forwarded to the Academic Senate's Committee on Budget and Interdepartmental Relations, a faculty committee that normally reviews tenure decisions. At that point, critics of the process, including Dean Richard Malkin, objected to the role played in the decision-making process by Professor Jasper Rine, a member of the UCB-N Advisory Committee (Abate 2003). The chancellor argued that no such conflict of interest existed and permitted Rine to remain on the committee. On June 5, 2003, the Budget Committee moved to deny tenure to Chapela. Despite some appeals by both the chair of ESPM and the dean of CNR, the Budget Committee made its decision final on November 20, 2003. The chancellor concurred with that decision.

Chapela appealed the tenure decision, and, apparently, a special faculty committee was convened to reexamine the dossier. Soon thereafter, Chapela also began legal action against the university, claiming that he had been denied tenure in large part owing to his opposition to UCB-N (Burress 2005). In addition, a protest rally by his supporters took place, and a petition signed by more than three hundred people, including university faculty from around the world, was handed to the UCB administration. Some time later the faculty committee recommended to the new chancellor, Robert Birgeneau, that Chapela be granted tenure (Lau 2005). Birgeneau concurred, and tenure was granted in May 2005. Chapela dropped the lawsuit against the university, even while vowing to keep up his fight against the biotechnology industry.

Regardless of the merits of Chapela's denial of tenure or subsequent reinstatement, there is little doubt that UCB-N played a role in it. First, the very existence of UCB-N changed the rules of the game. Certain faculty members were denied participation in the process because of the agreement. Second, while the administration saw fit to avoid conflicts of interest (COI) among faculty, they ignored the potential for COI among administrators. Thus, regardless of its validity, the decision of top administrators to accept the decision of the Budget Committee was seen by many as a COI. Finally, as a result of the conflict, the process of tenure review took far longer than is normally the case, leading Chapela to hold "office hours" in front of California Hall in protest.

Tyrone Hayes's Contract

While many people have expressed concern that UCB-N is qualitatively different from typical university-industry contracts because it involves almost every faculty member in one department, others have pointed with approval to this same feature as a safeguard against the disadvantages of contracts made between individual faculty members and a private company.

A panel discussion held at UCB on "the pulse of scientific freedom in the age of the biotech industry" presented the stories of four well-known physical scientists who had tussled with private biotechnology companies and felt that their experiences illustrated the influence that the biotechnology industry has over the work of public scientists (University of California, Berkeley 2003b). One of the participants was Tyrone Hayes, an associate professor of integrative biology, who was hired in 1998 by Ecorisk, Inc., to evaluate the effects of the herbicide atrazine on amphibians. The crop protection division of Syngenta gave $2 million to Ecorisk to fund these studies, and "in the contracts covering Dr. Hayes's work and that of many of the other researchers, Syngenta and Ecorisk retained final say over what and whether the scientists could publish" (Blumenstyk 2003). Hayes ran up against this provision when his research began to show that frogs were physiologically affected by low levels of atrazine. While working on this problem for Ecorisk, Hayes felt that the company inhibited and delayed his research and did what it could to prevent his publication of the results. The attempts to control the dissemination, and to question the credibility, of Hayes's work continued even after he left the employ of Ecorisk and began funding his research on atrazine himself.

This scenario points to the possible consequences for university-industry relations when the industrial partner is able to exert a seemingly disproportionate amount of power over the actions of the university partner. That such situations arise gives weight to the concerns of those wary of UCB-N with respect to how Novartis would try to influence the research directions and publications of the PMB scientists. However, this particular comparison also points to the potential differences between individual university-industry contracts and those involving entire departments. With greater numbers comes greater negotiating power. Certainly those public scientists acting in concert to create UCB-N had far fewer restrictions regarding their research than did Tyrone Hayes as a solitary public scientist. We shall return to this issue in a discussion of conflict of interest issues in a later chapter, but first let us turn to the debates surrounding the agreement itself.

FIVE

Points of Contention

W HILE THE IMPLEMENTATION of the agreement was relatively uncontested and while many of the critics' worst fears were unrealized, the fact that the agreement was widely challenged is important on a number of levels. The controversy over the agreement is informative in that it sheds light on some of the larger issues and contested transformations taking place in higher education in general and at UCB in particular. Furthermore, many of the controversies surrounding UCB-N are still having their effects, both locally and nationally.

The people we interviewed gave a number of reasons why the agreement was so controversial, which can be divided into four broad groups: (1) issues relating to the process by which the agreement was created and signed, (2) the substantive content of the agreement, (3) local conditions at UCB and in the Bay Area, and (4) broader issues that reflect the changing character of the university. Each group is briefly examined below.

PROCESS

The process by which UCB-N was developed and made public raised a very contentious issue. Indeed, much of the initial opposition to UCB-N was rooted in concerns regarding the process by which the agreement was formulated. For example, SRR, the graduate student organization that formed in response to UCB-N and remained active in subsequent debates concerning the future of

CNR, was first organized in response to the process by which the agreement was negotiated and presented.

Four aspects of that process were of concern to the people we interviewed. First, a number of faculty, students, and interested parties outside the university objected to the secrecy in which the agreement was formulated. Second, there was concern that the normal channels of governance and oversight for agreements with external parties were bypassed. Third, a number of participants argued that the way the agreement was presented and handled publicly was responsible for much of the controversy. Fourth, a number of faculty and graduate students raised questions concerning Rausser's consulting activities and his role in securing the agreement with Novartis.

Both proponents and opponents of the agreement generally agree that the process by which the agreement was formulated was less than transparent. While PMB faculty were kept abreast of negotiations, and while their input was solicited, other faculty in CNR and the university were not informed that negotiations were under way until the start of the semester in which the agreement was signed. Moreover, the contents of the agreement were not made public until after the official signing. A number of PMB faculty and UCB administrators argued that such secrecy was necessary in order to negotiate an agreement of this sort. One PMB faculty member remarked that the secretive character of negotiations was a "necessary evil," because the details had to be kept private in order to protect Novartis's business interests.

Other faculty and interested parties felt that they had a right to know what was in the contract and were entitled to a say in its outcome, because UCB is a public institution.[1] It is also likely that the secrecy would not have been an issue had the contract not encompassed (almost) a whole department. Those concerned with the lack of transparency were not interested in the details of (at least the vast majority of) industrial grants with individual faculty members but were concerned with an agreement of this sort of innovative and extensive scope.

The lack of transparency became especially contentious when the agreement was officially extended to CNR. Some faculty in the college outside PMB argued that they were included in the agreement because of the distribution of overhead to CNR as a whole but received no direct benefits as a result. Consequently, they argued, while they were burdened by the negative stigma attached to the agreement, they received no corresponding benefits to balance the situation. Regardless of whether they felt that secrecy was necessary, most interviewees agreed that the lack of transparency made the agreement more contentious.

In addition, critics argued that the agreement was negotiated in such a way that it bypassed the normal channels for agreements with external parties, particularly by going through OTL rather than SPO. Whether or not they thought this departure from normal protocol was justified, many faculty members believed that the agreement was deliberately developed in such a way as to preclude involvement by the larger university community. While the Academic Senate was involved, its participation was quite limited, as it was included only near the end of negotiations. Furthermore, the Senate was consulted in an ad hoc and highly contingent manner rather than in a way that tied together the workings of the administration, the Senate, SPO, and OTL. Indeed, while the involvement of the Academic Senate was not required, and while it could therefore have been bypassed altogether, the perception that the agreement was negotiated behind the scenes without following normal procedures eroded the principle of shared governance in the view of a number of faculty members. For faculty committed to the tradition of shared governance, having the administration and a single department—and moreover a new department—appear to go around normal channels made the agreement troublesome, regardless of its substance.

How the agreement was presented and communicated also became a source of problems. Many faculty, graduate students, and postdoctoral researchers within and outside PMB felt that they were unable to get adequate information about the agreement. One postdoctoral researcher in PMB remarked that, in his opinion, "the faculty at least, and the college, did a pretty poor job of keeping people informed and knowing what was happening and why it was happening ahead of time. They just presented this agreement with all its stipulations as a done deal and people jumped on the stipulations and didn't spend any time thinking, 'did it make any sense to do this sort of agreement?'"

In addition, as we saw in the previous chapter, critics charged that when information *was* made available about the pending agreement, it was done in a way that precluded discussion and debate. Rausser's "town hall meetings" were seen as a PR ploy designed "to convince everyone that this was a good idea." A faculty member in CNR said that these meetings were "so carefully stage-managed that before anyone said, 'and what do you people think?' the whole subtext was, 'never mind, it's a done deal.'"

The perception that information on UCB-N was being stage-managed was exacerbated by two events. The first was the hiring of a public relations person specifically to oversee the release of information about the agreement. Marie Felde, director of UCB media relations, and Robert Sanders, UCB senior public information representative, both noted that hiring an outside PR person was very unusual; neither could think of another occasion when this was done. The second event was an e-mail message from the CNR dean's office to all CNR faculty, suggesting that calls from the media be directed to public relations, which some faculty saw as an attempt to prevent criticism from reaching the media. Regardless of the dean's intentions, the e-mail made some CNR faculty even warier of the agreement.

Other actions taken by UCB faculty deepened the controversy as well. In 1988 Gordon Rausser and three other UCB faculty members had founded the Law and Economics Consulting Group (LECG). As allowed under UC policy, Rausser continued to be both a principal in LECG and a full-time Berkeley employee even as he became dean of CNR (1994–2000) and participated in the negotiations that created UCB-N. Although many benefits undoubtedly may be gained through such consulting agreements, people on and off campus were concerned about the arrangement between Rausser and LECG because of the large sum of money that Rausser reportedly made through his involvement and the ties LECG had with the forerunners of Novartis (Richardson 1997; Students for Responsible Research n.d.). In 1996 Rausser earned $1.3 million from LECG, significantly more than his salary as dean. Furthermore, the company held an initial public offering on December 18, 1997, at which Rausser planned to sell 16 percent of his shares for $2.2 million (Marshall 1997; Securities and Exchange Commission 1998). Finally, less than one year later LECG was acquired by the Metzler Group for $200 million.

These events raised a number of concerns about university-industry consulting arrangements in general and Rausser's involvement in particular. Len Richardson (1997, 5), editor of *California Farmer,* acknowledged that UC policy limits faculty members to fewer than forty-nine days per year of paid outside activity, but he asked in regard to LECG, "Can you build a $30 million business on 48 free days? Is the good name and reputation of UC becoming a faculty profit center?" He noted acerbically, "At least in the independent consultant community ethics is an issue that is recognized and debated, unlike in the University consulting community, University administration or regents" (Richardson 1998, 5). Others were also concerned that having the bulk of public university faculty members' income come from off-campus sources "compromises their academic objectivity" (Marshall 1997, C1).

There was also uneasiness about the close substantive ties between Rausser, LECG, and the Novartis parent companies. Along with other large agribusinesses like Monsanto and Heinz, both Sandoz and Ciba-Geigy were important clients of LECG, and in 1996 they merged to form Novartis. LECG offered its clients economic analyses on "environmental issues, water rights and technological innovation in agricultural machinery, hybrid seeds, engineered plants and species" (Securities and Exchange Commission 1998). People grew more disquieted about this arrangement as the UCB-N negotiations proceeded, with Rausser as a lead collaborator. As SRR put it, "The company

[LECG], and Mr. Rausser in particular, specialize in corporate mergers, IP, and agribusiness—just the skills that were improved in creating this alliance [UCB-N]. We feel there *may* be a conflict of interests" (Students for Responsible Research n.d.).

Several people inside and outside UCB mentioned Rausser's consulting activities in connection with the creation of UCB-N. As one CNR faculty member said, "Dean Rausser had been writing for years about how it was unavoidable, that universities had to get into these private relationships as institutions, these institutionalized relationships. So here he is presented with an opportunity." A number of interviewees expressed their belief that Rausser was using the establishment of UCB-N as an illustration of the concept of how public and private entities should interact in the future. Peter Rosset, co-director of Food First, and Miguel Altieri, a professor in ESPM, both pointed to this motivation for Rausser, especially given his history of being hired by the Consultative Group for International Agricultural Research and the World Bank as a consultant on building public-private partnerships in research institutions. That said, it is important to note that Rausser's interest in expanding industry-sponsored research in this time frame was widely shared by other university administrators across the United States (e.g., Condit and Pipes 1997).

Yet there was such an outcry regarding the involvement of UCB faculty members in LECG that UC president Richard Atkinson formed a task force to review UC policies on outside professional work (UCB Academic Senate 1998). In March 2003 faculty responsibilities were further clarified by the Technology Transfer Advisory Committee in "Guidance for Faculty and Other Academic Employees on Issues Related to Intellectual Property and Consulting" (UCB Office of Technology Transfer 2001).

In sum, from the perspective of those who favored the agreement, the process by which it was pursued was merely expedient. Something had to be done, and quickly. This was an expression of their autonomy both as individuals and collectively. To those affected negatively by the agreement, their lack of participation was evidence of hidden agendas and a lack of transparency at best, a direct threat to their autonomy at worst.

SUBSTANTIVE CONCERNS

In addition to concerns about the process, three parts of the agreement itself were particularly controversial. The first was its scale and scope. A number of the people we interviewed saw the agreement as problematic because it was with an entire department as opposed to a single faculty member or a small group of faculty. They argued that the scale of the agreement shifted responsibility from the individual to the institution and found this troubling, since if something went wrong, then UCB (or at least CNR) as a whole would be implicated. In addition, some argued that a department-wide agreement signaled an overly deep "penetration" of industry into the university and the partial or potential incorporation of the university into an "industry complex." Finally, an agreement with an entire department meant that the channels through which it would be negotiated, overseen, and executed were sufficiently underdeveloped that wider consultation ought to have occurred.

A related concern was the structure of intellectual property rights, a point to which we return in more detail later in this volume. The agreement, with some restrictions, granted Novartis the right of first negotiation on exclusive licenses to commercial research conducted by signers of the agreement—an option, in fact, to exercise that right on a portion of the results equal to the total percentage of external research funding that the company provided. Moreover, the right of first negotiation applied to all research conducted by signatories of the agreement, regardless of whether Novartis funded that research or not (with the exception of Department of Energy funding and funding from other private parties). In other words, if funds from Novartis constituted one-third of PMB external research funding, Novartis received the first right of negotiation to one-third of all PMB discoveries, including research funded by the National Institutes of Health, the National Science Foundation, and other public institutions.

Since Novartis only received the right of first negotiation, the license would still have to be negotiated to the satisfaction of OTL—though UCB would surely insist that licenses generate reasonable compensation. The company's expertise was such, however, that it could easily distinguish between significant inventions (i.e., dominating patent positions) and minor ones. It could not be assumed that the inventions remaining within the department or the university after Novartis's options were exercised would have a value proportional to their financial support. In short, critics expected that Novartis's deal would allow it to cherry pick all of the highest-value patents for licensing, leaving only the dregs for others. The principle here—that Novartis was given first right in the negotiation of patent licenses on research supported with *its* funds—has become routine at major research universities. But the fact that Novartis also received extensive access to publicly funded research caused widespread disbelief and outrage. Because Novartis would be given commercial access to such research, a number of critics argued that the agreement compromised the department and UC's autonomy, giving too much control and power to industry in the arena of patent licensing, even if it did not impede the autonomy of faculty in pursuing their own research directions.

Representatives from OTL disputed the claim that there was anything atypical about the agreement. They commented that the IPR arrangement, in addition to much of the rest of the agreement, was based on previous agreements between public research centers and industry. This defense nevertheless ignores the unprecedented nature of UCB-N in combining aspects of prior agreements. It was this uniqueness that led to the perception of UCB-N as a watershed in university relations and that heightened the controversy.

The third structural component of the agreement to cause consternation was the possibility that Novartis scientists would be given adjunct status, though, as it happens, this potential was never realized. Several of the people we interviewed believe that it did not because of strong opposition from faculty outside PMB. The intensity of this opposition can be seen in the results of the second survey conducted by ExCom. Few respondents to the survey thought adjunct status was appropriate for industry scientists from a firm that was providing funding to the college or one of its departments. A number of faculty members argued that granting Novartis scientists adjunct status would have bypassed the established governance procedures and stringent standards that are normally required for adjuncts. Many critics also felt that the offer of adjunct status was a way for Novartis to buy its way onto the UCB campus, while some supporters argued that this provision of the agreement was designed only to facilitate closer interaction between PMB and the company. One supporter argued that UCB faculty as a whole would never allow corporate researchers to become adjuncts and that everyone knew that. If this was true, then it is puzzling that the adjunct provision was on the table in the first place.

In addition to the particulars of the contract, the perception that UCB-N was unlike previous university agreements with industry also opened it to attack. While the novelty of the agreement is in fact debatable, nearly all of the people we interviewed agreed that an agreement with an entire department was something new. Its supporters tended to view the scope and scale of the agreement as unproblematic. A number argued that in fact its department-wide nature would allow for greater academic freedom and enable more faculty to participate.

A number of supporters thought that much of the controversy was the result of a poor understanding among faculty of UIR. They argued that faculty in the social sciences and humanities knew little about "normal research policy" or university relations with external parties. Paul Gray, the EVCP of UCB, remarked, "It was a kind of a case where the arts and humanities and social science side of campus all of a sudden were brought face to face almost by accident with this animal they hadn't really seen before and weren't really familiar with." It was also argued that faculty in CNR tended to have little experience with industry agreements, whereas faculty in engineering had been conducting relations with industry for a long time. (At the same time, this argument brackets the historical emergence—within computer engineering at least—of a situation where patents and licenses have come to be seen as an impediment to creativity rather than an inducement.) Because faculty in CNR and other departments were unaware of the history of UCB's relations with industry, some interviewees argued, UCB-N was blown out of proportion.

In particular, proponents singled out IPR as an area that many opponents of the agreement did not understand; they argued that IPR thereby became the source of more controversy than should have been the case. A number of people also accused opponents of the agreement of criticizing it without knowing its specifics. As one faculty member in PMB remarked, "They comment on what they believe to be the content of the document ... rather than feeling compelled to read the document before making comment." (Here, critics such as the graduate students in SRR would argue that, try as they might, they couldn't get a straight answer as to the actual content.)

In short, the very provisions of the agreement that proponents saw as desirable, or at least as reasonable, detractors saw as compromising. Supporters saw the department-wide access to faculty, the new (or converted) building, and the adjunct status for Novartis researchers as enhancing the potential for creativity and for cementing PMB's role as a leader in the field. The generous IP options were therefore a reasonable enticement to offer Novartis. Opponents, for their part, saw the agreement as selling off the university to the highest bidder, and, at least potentially, as compromising academic autonomy, limiting research creativity, and committing the administration to constraints on intellectual diversity.

LOCAL CONDITIONS

The historical legacies and local conditions of UCB and the Bay Area were significant contributing factors to the controversy and its magnitude. A number of the people we interviewed commented that the agreement generated so much controversy because it occurred at UCB, which many consider one of the most liberal campuses in the United States (a point disputed by others). Moreover, as Dave Henson, the director of Occidental Arts and Ecology, a nonprofit educational organization, argued, the Bay Area community "has the largest active set of proponents of ecological agriculture and opponents, for very good and diverse reasons, of corporate privatization of genetic engineering." Some interviewees believed that if such an agreement was going to arouse criticism, it would happen at Berkeley. Furthermore, because of its reputation as both a leading university and the site of much social activism, it was suggested that Berkeley also attracts more public interest and media attention than most universities. If the agreement with Novartis had involved another UC campus, they claimed—Riverside, say, or Davis (the other UC campuses where research on conventional crop plants is conducted)—there would not have been nearly as much public scrutiny or media attention (see, for example, Knudson and Lee 2004).[2] This is speculation, of course, but one should remember that the formation of Calgene in 1980 by UC Davis professor Ray Valentine made national headlines, and that there was deep statewide resistance to field trials of the antifreeze tomato developed at UC Davis (Bird 1993).

At the same time, within UCB, and most notably in CNR, historic relations and local conditions very much influenced how UCB-N was perceived. As noted earlier, both populist and progressive (often termed "productivist") sciences have historically been pursued in CNR. Productivist scientists are generally interested in making agriculture more productive, while populist scientists are interested in the needs of small farmers and the social and environmental impacts of large-scale agriculture. While both have coexisted in CNR, there have always been tensions between the two, and these tensions have been exacerbated in the past decade owing to changes made in CNR. One overview of agricultural pursuits on campus dating back to the 1970s concluded, "Although some departments continued work relating to agriculture, its overall focus shifted increasingly toward conservation and resource studies and toward ever more fundamental and theoretical work in the natural sciences. In Berkeley's highly urbanized, intellectual milieu, relatively little understanding or sympathy remained for production agriculture, and in fact some of the most stinging criticisms of California's agricultural system would eventually emerge from there" (Scheuring 1995, 176).

Many of the people we interviewed argued that the two sides are culturally and philosophically opposed and that this opposition was responsible for much of the controversy. Nevertheless, as is often the case in such matters, each side tended to depict the other as monolithic and without nuance. It can be argued that much of the concern over UCB-N was the outcome of the long-standing divides among some faculty in CNR.

One such divide is between what one interviewee termed the "old biology" (botany, zoology, and so on) and "biotechnology." This division is most evident in the different perspectives and approaches taken by ESPM and PMB. A number of interviewees pointed to the inherent strife between these two departments, which are interested in similar problems but approach them from very different standpoints. As one faculty member put it, "In the case of plant biology, so far as I can perceive, you study the natural world to bend it to human interests and needs; whereas in botany, you study the natural world, plants of course, to help humans fit into a broader fabric of nature." This philosophical difference has been a long-standing divide in CNR, but a number of CNR faculty believe that UCB-N acted as a lightning rod. This may have been because, as one interviewee observed, UCB-N signaled that PMB and molecular approaches were "winning," while other approaches to biology and ecology were becoming marginalized.

BROADER ISSUES

The broader concerns that became intertwined with UCB-N revolved around the appropriate role of industry in a public university and the kinds of research it should undertake. These concerns reflect a more general crisis and set of transformations that are taking place in higher education throughout the United States regarding what is public, what is private, and the appropriate relation between the two. Much of the opposition to UCB-N was rooted more generally in debates over the "privatization" of things that were once public—whether ecological, personal, cultural, or spatial. For many critics of the agreement, industry involvement with public universities, and with biotechnology research in particular, represents the increasing encroachment of the private on the public sphere.

A number of those interviewed asked whether a corporation had any appropriate role in a public university. One asked, what is "the rightness, appropriateness, the morality of a private company sponsoring research at a university using facilities paid for with tax money for the company's eventual profit?" Others raised concern about whether Novartis, because it made up a significant percentage of PMB funding—28 percent in fiscal year 1999—would directly or indirectly control the kind of research undertaken by faculty.[3] There was also concern over whether Novartis was using the university as a (relatively) cheap source of labor. Critics of UCB-N accused UCB of "giving away our patrimony ... to Novartis for chump change." In some people's minds Novartis was given free rein to take and privatize research that had been made possible by decades of public investment. Although new to many UCB faculty, this issue is hardly new in higher education (see, for example, Fairweather 1988).

A number of people we interviewed were also worried that conflicts of interest would develop as a result of university relations with industry. These worries took two forms. On the one hand, some questioned whether faculty funded by Novartis would be able to fulfill both their private, contractual obligations to Novartis and their public, professional responsibilities as members of a public university. On the other hand, a number of faculty, particularly those in ESPM, were concerned about whether aligning CNR with Novartis would undermine their ability to participate in policy debates and to be perceived as objective experts.

Concerns regarding the kind of research that a public university should undertake arose primarily in response to questions about the appropriateness of biotechnology research at UCB. At the time that UCB-N was being debated there was escalating disagreement over agricultural biotechnology in general. The development of agricultural biotechnologies and the concurrent expansion of the industry have been paralleled by a "proliferation of citizens' voices challenging the biotechnology industry on economic, environmental, cultural, and moral grounds" (Schurman and Munro 2003, 111). Without question, debates over agricultural biotechnology influenced how both supporters and critics of the agreement understood it and the controversy that surrounded it.

For many critics of the agreement, research on biotechnology goes against the land grant and public missions of the university because it largely serves international agribusiness corporations rather than the people of California. It is difficult to evaluate how residents of the state could benefit directly from UCB-N, given Novartis's home in Switzerland. Indirect benefits might otherwise be measured, critics feel, against those that alternative research agendas—biotechnological and otherwise—might have provided. For many populist agricultural and environmental advocacy organizations, opposition to biotechnology is grounded in a critique of the kinds of biotechnologies that are currently being developed rather than an opposition to biotechnology in and of itself. Peter Rosset, co-director of Food First, argued that CNR "shouldn't be developing products for private profit but it should be looking at what are the social and ecological impacts of various ways of exploitation of natural resources and what are alternative ways to exploit them in a more sustainable fashion." Several faculty and graduate students in CNR made the parallel argument that the increasing focus on biotechnology research is forcing other approaches to agriculture and ecology to the margins and reducing the diversity of science within the college.

In interviews, a number of supporters of the agreement argued that ideological opposition, rather than any substantive issues regarding the agreement itself, was responsible for most of the controversy. Some interviewees argued that UCB-N fed into and brought together existing opposition to both biotechnology and corporatization. A central administrator remarked that UCB-N was so explosive because it brought together biotechnology and globalization in ways that other agreements with industry, such as those in engineering, had not previously done. However, while both proponents and opponents agreed that UCB-N became so controversial in part because it was with Novartis, a multinational biotechnology corporation, very few opponents of the agreement named biotechnology as among their chief concerns. In other words, while concerns regarding biotechnology clearly contributed to the controversy, supporters of the agreement tended to see it as a larger concern than it actually was for most opponents.

CONCLUSION

The controversy that surrounded the agreement did have some tangible effects. While it is not possible to establish clear causation, the level of concern appears to have affected the implementation of the agreement. Novartis scientists were not granted adjunct status and the interface facility was not constructed. The controversy may have discouraged the UCB administration, other public universities, and corporations from entering into similar kinds of agreements. The controversy also seems to have had some lasting effects within CNR, as illustrated by the ongoing debate on academic diversity. Moreover, it continues to affect certain individuals. This is most clear in the conflicting perceptions of the reasons for the delays and reversals in Ignacio Chapela's tenure process— and its relationship to the debate on academic diversity. From the perspective of Novartis, while the controversy was quite surprising and created some negative public relations, it has had few, if any, lasting effects. One representative of the company remarked that the agreement was "definitely not on the radar screen of the shareholders," and that most of their customers would not have heard of it because such a small amount of money was involved. Finally, it should be noted that the controversy did produce an external review, and it increased the possibility for dialogue on the future of the university. We now turn from the perceptions of proponents and opponents to the agreement itself. We set it in the context of other university-industry relationships, and we look at what it contained and how it was implemented.

SIX

Overview and Analysis
of the Agreement

A S DETAILED BY ZUCKER, Darby, and Armstrong (2002), commercially viable innovations may be expedited when academic and industrial scientists work closely together. Legal agreements between universities and commercial entities often define these relationships and associated financial arrangements. Numerous tutorials instruct practitioners in how to bridge the gap between university and business practices when drafting agreements of this type (American Council on Education and the National Alliance of Business 2001; Berneman 1995). UCB-N is one notable illustration of the complexities and challenges that can emerge in the process. With its ten appendices, the agreement runs sixty pages. It defines the parties' joint interest in conducting basic science and NADI's interest in developing commercial plant traits. During its brief history, in collaboration with PMB, NADI specialized in plant genomics with special attention to *Arabidopsis* and rice.[1] The agreement terms and conditions detail the IPR of both parties.

For the academic community, UCB-N stood out because it represented significant industrial rather than government funding for Berkeley researchers. PMB's financial objective in making the agreement was to secure a sponsored research agreement from an industrial sponsor of at least $5 million a year over five years. The effort's success at winning industrial patronage on this scale was unique in UCB's history. Of the twenty-six awards of $5 million or more received by UCB in fiscal years 1998 through 2003, only four were not from the federal government. Two of these came from foundations and one from the state government; the final one was UCB-N.[2] Thus the continuing importance of sponsored research projects from federal agencies can be seen in the projects received by UCB over the term of UCB-N. At least for the life sciences, the bleak forecast for federal support that had informed a president's retreat less than two years earlier had turned bright, as significant new and multiyear funding began to flow into genomics.[3] For example, UCB's Molecular and Cell Biology Department (MCB) received $38.6 million from the National Center for Human Genome Research in FY 1998. In 2002 UCB launched the Center for Integrated Genomics, which includes the Departments of MCB, Integrative Biology, Statistics, Computer Science, Bioengineering, Plant and Microbial Biology, Biostatistics, Mathematics, Physics, and Public Health, and the Lawrence Berkeley National Laboratory.

Of particular relevance to PMB, the most significant opportunity for extramural support at the time was the collaborative research and infrastructure projects within NSF's plant genome research program. PMB's assertive effort to scale up its capacities in genomics was part of a larger competition for newly available federal funds, as well as the result of industrial interest in the field. At the same time that UCB/CNR/PMB were negotiating their agreement with Novartis, Cornell University assembled a genomics task force that defined the Cornell Genomics Initiative, Purdue University launched its Agricultural Genomics Initiative, and the Donald Danforth Plant Science Center in St. Louis, Missouri, was established as an independent nonprofit. Funding for the Danforth Center came from the Danforth Foundation and the Monsanto Fund to support researchers from the University of Illinois, the Missouri Botanical Garden, the University of Missouri (Columbia), Purdue University, and Washington University (St. Louis) to work alongside Monsanto scientists. Berkeley researchers and administrators proved successful in the subsequent competition for federal funds. Of the twenty NSF genome program awards of more than $5 million, UC was the lead institution on five and a partner on three others.

Proper analysis of UCB-N relative to the more familiar support of the federal government requires sensitivity to the contested state of UIR norms at the time of the agreement. Attempts to commercialize biotechnology in the 1970s and 1980s exacerbated existing populist-progressive tensions within public universities. From the first patent issued in the field, the demarcation between academia and commerce fostered disagreement (Hughes 2001; U.S. Senate Committee on Commerce, Science, and Transportation 1978). By 1981 commercial interests were "seriously dividing members of the academic community" (Fox 1981, 39). A few years later, the U.S. Congress's Office of Technology Assessment (OTA) examined industry-university relationships within land grant universities. OTA documented concerns "over who controls the university research agenda, the allegiance of scientists to their university employer, the willingness of scientists to discuss research discoveries related to potentially patentable products, and potential favoritism shown particular companies by the university because of its research ties" (U.S. Congress, Office of Technology Assessment 1985, 71).

While such questions lingered, academic entrepreneurship expanded markedly in California, and with lasting results. By the 1980s leading life scientists had garnered considerable business experience. Prominent examples include UCB professor Donald Glaser's founding of the Cetus Corporation (1971); UC San Francisco professor Herbert W. Boyer and Stanley N. Cohen of the Stanford University School of Medicine launching Genentech (1976); UC Davis professor Ray Valentine and Stanford's Roger Salquist establishing Calgene (1980); UCB professor Edward E. Penhoet[4] co-founding Chiron Corporation along with his former professor, William J. Rutter, and Pablo Valenzuela (1981); and UC Davis professor Raymond Rodriguez's launching of Applied Phytologics, Inc. (1983). Many observers saw this trend as proof that leading scientists could work effectively with industry *and* in the public interest. Reflecting on his own industrial experience with a startup company (DNAZ) and a multinational firm, Stanford Nobelist Arthur Kornberg wrote:

> For the university concerned with pursuit of knowledge and its transfer to promote human and economic welfare, the DNAZ achievements demonstrate that basic science can be pursued quite effectively in another setting, and they should direct attention to the shortcomings inherent in the operations of large academic institutions that impede the creativity of their faculties. Finally, the DNAZ-Schering-Plough relationship illustrates that, in the long term, the scientist and the industrialist, with mutual confidence and trust, can produce good science and sound business of significant social merit. (Kornberg 1995, 260)

In addition to startup ventures, early university-industry agreements in biotechnology included Monsanto and Harvard Medical School's $23 million twelve-year deal beginning in 1974 (Culliton 1977); Harvard Medial School's affiliate Massachusetts General Hospital's $70 million agreement with Hoechst AG in 1981 (Bowie 1994), which gave Hoechst the exclusive right to the commercialization of any invention from Mass General's Department of Molecular Biology; Allied Corporation and UC Davis's $2.5 million agreement in 1981; Washington University Medical School and Monsanto's $23.5 million five-year 1982 agreement (Kenney 1986); and UC Irvine's Department of Biological Chemistry collaboration with Hitachi Chemical Company on a $12 million deal in 1988. In the last case, industrial and academic researchers shared the same building (National Research Council Committee on Japan 1992). Other agreements of note included DuPont's collaborations with the California Institute of Technology, Harvard University, and the University of Maryland; and W. R. Grace's work with MIT, Rockefeller University, and Washington University (U.S. Department of Commerce, International Trade Administration 1984).[5]

Numerous concerns simmered as the number of university-industry agreements grew. Points of contention included the potential loss of U.S. competitiveness in biotechnology to foreign firms teaming with American scientists, LGUs being left at a disadvantage relative to private universities, and industry in greater control of research agendas (U.S. Congress, Office of Technology Assessment 1991; Blumenthal et al. 1986a, Blumenthal et al. 1986b). A government-university-industry research roundtable organized by the National Academy of Sciences stepped in to codify acceptable terms and conditions for sponsored research agreements in 1988. The roundtable further amplified workable approaches to the still contentious management of IP in 1993 (National Academy of Science Government-University-Industry Research Roundtable 1988, 1993).

With sporadic success stories and occasional role confusion spreading across American campuses, a number of new government policies advanced the academic role in innovation and economic development. Responding to the economic stresses of the 1980s and 1990s, lawmakers passed a number of government initiatives aimed at removing perceived restrictions on American innovation. Policymakers combined the domestic cold war scientific agenda with a global industrial competitiveness rationale to loosen antitrust laws, expand IP practices, relax financial restrictions, and weaken environmental regulations. The flow of legislation included the following acts of Congress:

> Stevenson-Wydler Technology Innovation Act (1980)
> Bayh-Dole University and Small Business Patent Act (1980)
> Small Business Innovation Development Act (1982)
> National Cooperative Research Act (1984)
> Federal Technology Transfer Act (1986)
> Omnibus Trade and Competitiveness Act (1988)
> National Competitiveness Technology Transfer Act (1989)
> National Defense Authorization Act (1991, 1993, 1994, and 1995)
> National Technology Preeminence Act (1991)
> Defense Conversion, Reinvestment, and Transition Assistance
>   Act (1992)
> Small Business Technology Transfer Act (1992).
> Small Business Research and Development Enhancement Act (1992)

Private Securities Litigation Reform Act (1995)
Small Business Programs Improvement Act (1996)
Small Business Reauthorization Act (1997)[6]

Universities and corporations found new ways to work together using independent boundary organizations to help relieve the tension between the private and public sectors.[7] In this way, universities could claim an "arm's-length" relationship with for-profit aims. This sort of quasi-public/quasi-private research organization had been pioneered by the Research Corporation, established with the assistance of UC in 1912, followed fairly soon thereafter by the Wisconsin Alumni Research Foundation, formed in 1925 (Marcy 1978; Apple 1989). Over the years, the university-industry boundary-organization approach evolved, taking on many names with essentially the same function (e.g., industrial parks, research parks, technology parks, business technology centers, and business incubators). Brooks and Randazzese (1999, 383–85) illustrate the nature and function of boundary (or buffer) organizations:

> The incompatibilities between industry and the academy—those of research style, objectives, disclosure policies, intellectual property, and conflicting financial interest—threaten to undermine the basic purpose of universities and frustrate the expectations of the economic benefit that comes from closer university cooperation with industry. ... Many communities, often with state government assistance, have created incubator institutions next to universities to encourage the formation of new firms based on university research. The core employees of the buffer institutions could carry out the necessary translation of academic research results to the point where they could be more readily adopted by small and medium companies.

The recent proliferation of these arrangements led to two new professional organizations. The National Business Incubation Association, a membership organization of incubator developers and managers, was established in 1985, and the Association of University Research Parks was launched in 1986.

Other forms of boundary organizations (including professional and trade associations, as well as agricultural commodity groups) developed during the twentieth century and grew to become important vehicles for the formulation of government standards, regulations, and trade policy. By the 1990s these types of organizations were specifically empowered by government. For example, the National Technology Transfer and Advancement Act of 1995 states: "Federal agencies and departments shall consult with voluntary, private sector, consensus standards bodies and shall, when such participation is in the public interest and is compatible with agency and department missions, authorities, priorities, and budget resources, participate with such bodies in the development of standards" (P.L. 104-113, section 12d).

The winding down of the cold war, the increasing competitiveness of ever more global markets, and the declining relative efficiency of American industry raised concerns that America's science infrastructure would erode under declining federal funding. NSF noted the changing climate: "As the 1990s opened, the United States faced the novel challenge of redefining its goals and priorities in the post-Cold War era" (National Science Foundation 2000, 1-21). This reprioritizing shifted attention from national security questions between two stable superpowers to economic development questions associated with the instabilities of international markets (Kleinman and Vallas 2001; Slaughter and Rhoades 1996; Brooks and Randazzese 1999). MIT president Charles M. Vest (1996, 71) captured the pressure for change concisely when he wrote, "We cannot justify the investment of federal funds in 1996 on the fact that we won the war in 1945. We need to look to the future, conceive of improvements in our role in the national innovation system."

In this political and economic climate, state universities rediscovered a vested interest in championing, or at least claiming credit for, economic growth in order to show a return on government support (Eisinger 1988). Yet Osborn (1988, 37) argued that a barrier to economic vitality was "the chasm that separates academia, where most basic research is done, and business, where new products are created." University administrators and scientists alike were generally reluctant to bridge the chasm, but they ultimately awoke to the political reality that concern for the economy was the new "ticket to survival and to resources" (Kaplan 1986, 100). Reflecting the tension, UC guidelines released in 1989 made clear the external nature of the new agenda by stating that not the university but "legislators increasingly see [university-industry] cooperation as a way of enhancing national R&D efforts and of helping to make the State and the nation more competitive" (University of California, Office of the President 1989). LGU faculty involved in agricultural research also saw the need to align with the new agenda, one decidedly different in its focus from the emphasis on rural and community development found in the Smith-Lever Act, especially as amended in 1955 (Feller 1987; Feller et al. 1987).

In general, the terms and conditions of UCB-N appear consistent with the behavior of universities adjusting to the emerging norms of university-based economic development. But given the history of biotechnology, it is not so surprising that the agreement drew immediate fire. After the squabbles that began in the 1960s about nonproliferation, laboratory safety, IPR, industrial concentration, academic entrepreneurship, field trials, environmental protection, and food regulations, one might reasonably view heated debate as the norm. In other words, it cannot be assumed that controversy was solely the result of the agreement's content or construction. The heart of the issue is the coincident intellectual and institutional restructuring of the biological sciences alongside the restructuring of universities and university-industry relations. Neither proponents nor opponents of UCB-N appeared aware of the coincidence of these two conceptually and practically distinct phenomena, yet this coincidence proved to be a central but unspoken feature of the UCB-N controversy.

In fact, UC behavior regarding agricultural biotechnology in general appears to be in step with government policy, if somewhat out of step with the concerns of federal scientific boards. UCB-N, properly labeled an experiment, also falls neatly into the learning-by-doing evolution of university technology transfer. Perhaps the closest comparable agreements in time frame and subject matter are those used by the Consortium of Plant Biotechnology Research, Inc. While UC did not elect to participate, the consortium counted most of the key LGUs and biotechnology firms as members. The standard research agreement of this largely federally funded group did not deviate markedly from the UCB agreement in grant of rights or restrictions on IP. It is important to note, however, that consistency with the aims advocated by governmental actors does not eliminate the need for deliberate discussion of the appropriateness and desirability of such arrangements.

As we have seen, UCB-N was not far from the norm in the late 1990s, yet it did have unique characteristics. One key deviation from the norm was the inclusion of nearly an entire academic department in an agreement with a single firm. UCB administrators knew that the "collaboration by a large number of faculty with a single corporate entity over five years [was] unique" (UCB Public Information Office 1998). Such an approach lay outside the mainstream. Accordingly, in an informational booklet on university technology transfer, the Council of Governmental Relations (1993, 20) asked the salient question: "When is it appropriate for license rights to future federally assisted inventions to be committed to an industrial sponsor?" The organization of research universities answered: "universities should not grant to a single industrial sponsor the rights to federally assisted inventions from the entire institution or major units such as departments, centers and laboratories. The granting of rights must be specific to the scope of work funded."

Also relevant here was the controversy surrounding a long-term multimillion-dollar agreement between Sandoz Pharmaceutical Corp. (later part of Novartis AG) and the nonprofit Scripps Research Institute in La Jolla, California. This 1992 agreement may or may not have generated the Council of Governmental Relations booklet, but it certainly precipitated new guidelines from the NIH. These guidelines cautioned NIH grantees when "the sponsor's prospective licensing rights cover all technologies developed by a major group or component of the Grantee organization, such as a large laboratory, department or center, or the technologies in question represent a substantial proportion of the anticipated intellectual output of the Grantee's research staff." The NIH advised that a "Grantee should consider establishing some mechanism to limit the review and licensing rights of the sponsor to a particular segment or percentage of the inventions and for a set period of time" (National Institutes of Health 1994, 13). UCB-N fell short of at least the spirit of NIH's guidelines, since all noncommercial inventions were open to review by the sponsor.

A second important deviation from the norm was the extended capture of government-funded IPR. Precisely what inventions fell under the agreement and whether the IPR arrangements covered the whole department calls for further elaboration. The agreement was a legal contract between two parties, namely, NADI and UC (in this case, technically, the Regents of UC). As is normal practice, it was not a contract with a department or college but with UC as a legal entity.[8] PMB as a department is mentioned in the agreement regarding oversight of the research and advisory committees, but the agreement remains with UC. The inventions within the scope of the agreement are tied to the specific individuals who signed on to the agreement and not to PMB as a whole. About thirty PMB faculty members signed on over the term of the agreement, but a few did not. As such, the inventions of the whole department did not fall under the agreement. However, a collaborative project involving one or more of the few not signing and anyone receiving funds under the agreement would have fallen under the agreement. In sum, it is fair to say that the whole department was affected by the agreement. At the same time, one could argue that UC stayed within the letter of the NIH guidelines, since not all members of PMB signed UCB-N, though they couldn't have known this during the negotiations.

Under the agreement, the university owns the IPR developed and NADI has the first right to negotiate a license on a portion of the patentable discoveries made. While the Bayh-Dole Act provides for a small business preference, it includes a provision that a right of first refusal may be extended to a large firm in consideration for multiple-year funding. NADI could elect to negotiate an exclusive license within a set range of royalties when an invention led to a seed trait, or without predetermined royalties if the invention did not. NADI received a free nonexclusive right to inventions that derived in part from its proprietary information. Defining royalty rates in a research agreement is rare and suggests the expectation that valuable seed traits would be generated under the agreement.[9] With a pre-negotiated range, a commercial entity is assured that royalties will not exceed a given ceiling, while UC has a floor below which royalties cannot fall. To fall under the agreement, inventions by researchers needed to be funded by NADI or a noncommercial entity such as a government agency (e.g., NIH, NSF, USDA).

At the same time, the agreement is carefully crafted to acknowledge UC's real and potential constraints under its policies, as well as federal laws and contractual obligations with third parties (e.g., government agencies and commodity groups). For example, the language "to the extent the University is legally able to do so" hints at possible entanglements or restrictions regarding control of future inventions created with public funds. One example of the agreement in practice is a U.S. patent application for a gene sequence and promoter.[10] This application derived from research conducted with both USDA and TMRI funds by PMB researchers. Under the terms of the agreement, the benefit of the government funding flowed directly to a single firm. This arrangement strains conventional thinking on the proper stewardship of public funds.

The conclusions of the National Science and Technology Council's Interagency Working Group for Plant Genomics provide a useful baseline for other concerns about UCB-N. Ten months before the agreement was signed, the interagency working group released its final report, in which it sought to address IPR concerns in the field but did not have a definitive approach for resolving public-private tensions:

> Limitations in the availability of public funds ... combined with current patenting policies, have led to a significant investment by the private sector which now is unable to make the information freely available. While it is undesirable to duplicate private sector efforts, both philosophically and economically, the government must now act to provide critical data and research tools to the entire plant science community. Government officials, as representatives of the public research community, should continue to hold discussions with private industry in an effort to minimize current and future impediments to plant genome research.

Moreover, the government report recognized industry's coincident interests and was sensitive to the private sector's position:

> [C]ompanies have become major players in all phases of genomics, including plant genomics, with a particular focus on crops with the highest market value such as corn and soybeans, in the United States. Biotechnology

companies face the challenge of a long product development cycle, making them particularly dependent on the ability to obtain enduring capital investment through the research, development, and manufacturing period preceding product marketing. (National Science and Technology Council, Committee on Science, Interagency Working Group on Plant Genomics 1998)

Although UCB-N covered some questionable terrain, it seems unreasonable to expect UC faculty or administrators to have found a timely and elegant solution to difficulties that perplexed the federal government. At the same time, it is striking that in this context these national concerns appear to have generated few reservations on the part of PMB, Rausser, and other members of the UCB administration, as they moved forward in negotiating the agreement.

Finally, a persistent theme in the debate over the ideals and norms of academia rests on the premise that government funding can be roughly equated with basic or pure research, while industrial funding is equated with applied research tied to potentially suspicious motives. The ideals of basic science, in turn, underlie the concern that academic research agendas may creep undesirably toward the interests of industry. This sort of analysis assumes a short-term vision on the part of industry—a vision at odds with Novartis's Bell Labs-like commitments to NADI at the time of the agreement—and fails to take into account the extensive history of federal funding of applied research projects so prevalent throughout the history of agricultural research funding.

Furthermore, one consequence of the increasingly global world of neoliberal political and economic governance has been the institutionalization of new and deeply held beliefs that universities are engines of economic growth, and this consideration, too, had an impact on the crafting and reception of UCB-N. In this view, the interests of the private sector are not significantly different from those of the public. The dominant trajectory of the life sciences today—as evidenced by the intended scope of the agreement—holds that government, research scientists, academic administrations, and industrial entrepreneurs and corporations have effectively coincident interests relative to the science at hand (e.g., National Science Foundation, Office of Legislative and Public Affairs 2001). In this context, criticism about potential changes to or constraints on PMB's research agenda look misplaced. At the same time and for the same reasons, claims by PMB faculty that their research agendas would not be, and were not, altered by the relationship with Novartis appear questionable.

In sum, the period before and during the agreement was marked by a growing consensus within university administrations, government, and the private sector that closer relations between universities and the private sector were desirable, if not an unqualified good. Although there were certainly critics and detractors, few raised the tough questions about the role and mission of public and land grant universities; few raised questions about the consequences of choosing this path for the autonomy, creativity, and diversity of the college and university as a whole. While it would be unfair to say that the proponents of closer relations between universities and industry ignored the public good, broader questions of that sort were often brushed aside as of merely "philosophical" import. What mattered were the immediate goals of securing financing, advancing scientific research, and balancing annual budgets.

But if those who crafted the agreement were largely attuned to legal nuances and government policies, the same was not true of the press, which played a considerable role not only in reporting the events surrounding UCB-N but also in shaping those events, as we shall see in the next chapter.

SEVEN

The Agreement and the Public Stage

T HE AGREEMENT BETWEEN the University of California, Berkeley and Novartis became a public concern on October 9, 1998, when the *San Francisco Chronicle* reported that these two organizations were in the final stages of their negotiations. The article mentioned that similar agreements between universities and private corporations had become commonplace, but it raised concerns that the scope of this particular arrangement would give industry an unusual degree of control over the research agenda on the Berkeley campus. On October 14 a letter to the editor of the newspaper echoing these concerns was the first sally in what very quickly became a public debate.

These two newspaper items, which appeared prior to the actual signing of UCB-N, shed light on the role of the media in shaping public opinion. Obviously, the media report on issues as unproblematic as the weather or the outcome of a ballgame, and as contentious and inflammatory as agricultural biotechnology and Supreme Court decisions. When readers and viewers see coverage of an issue in which they have a vested interest, they may mobilize resources to try and influence that coverage and thus public opinion (e.g., Montgomery 1989). This was certainly the case with UCB-N, and this chapter focuses on what was said in the public arena, and by whom.

It should be noted at the outset that it is not our intention to review the operational guidelines of various news organizations, which have been covered by others (e.g., Bagdikian 1992; Franklin and Murphy 1998; Gans 1979; Gitlin 1980; Tuchman 1978; Zelizer 1992), but to show how the media bring to light discrepancies or disagreements over how other organizations operate.

We analyze both internal and external press coverage of the agreement to show how the principles discussed in Chapter 2 (creativity, autonomy, and diversity) were upheld by media/public relations offices within the university and challenged by media organizations off campus.

THE ROLES OF THE MEDIA

Our approach is grounded in the social-problems model put forth by Blumer (1971) and Hilgartner and Bosk (1988), which defines a social problem as something that appears in a public arena, such as the mass media, which leads to public awareness and controversy in some form or another. The publicity then generates various forms of legitimation processes by which those claiming to be knowledgeable about the issue at hand step forward to offer their suggestions and possible assistance in solving the problem. Researchers then measure, operationalize, and analyze the discourse that takes place in the public arena in order to gain an understanding of how the controversy has been constructed and how its construction is interpreted and responded to by different sectors of the audience (Ader 1995; Best 1991; Lange 1993).

In addition, there is a "bounce" or "ripple" effect when actors with a vested interest in the issue, whether directly or indirectly, react to the various constructions coming before the audience. Those with direct ties may find that they have to try to justify their conduct, while those with indirect ties may change the way they conduct themselves in the future so as to avoid similar problems and controversies. Both media effects are important for understanding the impact of UCB-N, as many different actors were (potentially) affected by it. This became clear in a number of articles appearing in such publications as the *Chronicle of Higher Education,* in which individuals with a vested interest in the image of higher education questioned various aspects of the agreement.

Our framing of UCB-N as a social problem stems from the fact that coverage of it included confrontations between UCB administrators and people who felt that UCB-N threatened UCB's academic freedom. We have looked at some of the opposition to the agreement in earlier chapters. In addition, some reporters were interested in the agreement purely from a news standpoint.

Once news reporters decide that an issue is controversial, professional ethics dictate that opposing viewpoints be presented, though equal coverage is rarely given to all sides of a question (Gans 1979; Gitlin 1980; Tuchman 1978). Ericson et al. (1989) argue that big business, the criminal justice system, and government offices have more control over media coverage than smaller, less powerful interests. These organizations have more direct links to reporters, which gives them opportunities to shape the news. In contrast, smaller groups are often overlooked, underplayed, or reinterpreted as reporters and editors add their own take to the stories coming from such sources, often to the benefit of their "cash cow" sources. Given the resources controlled by UCB and Novartis, we would expect these organizations to be able to have some control over the content of news reports related to their collaborative arrangement.

REPORTING A PROBLEM

When (real or perceived) norms—including the kinds of conventions discussed in Chapter 2—are breached, conflicts can arise within organizations (Morrill 1991) or between individuals (Schonbach 1990). When the conflict becomes noteworthy, either because of the actors involved (e.g., celebrities, politicians) or the degree of escalation (e.g., armed conflict), the media are usually expected to report on it (Ten Eyck and Deseran 2004). Within an institution such as UCB, where many people are considered experts because of their training and titles, determining who is the best authority on such issues as academic freedom and creativity becomes problematic. In our risk society (Beck 1999), many people look to scientists and politicians to offer advice on various threatening issues, among them the impact of industry finances on university research agendas, as most people do not have the time or resources to study and understand these topics. The media often play a key role when questions arise over the legitimacy of these authority/expert figures (Thompson 1990).

In such a situation, the accuracy and integrity of reporters is often questioned, given that reporters are an interpretive community (Zelizer 1992, 1993). This simply means that they follow the practices of those who have been successful, while attempting to avoid mistakes for which others have been punished. Given the amount of coverage given to such issues as the Vietnam War protests (and the central role of UCB in those protests), higher education, and biotechnology—and the number of accolades and awards bestowed on reporters who covered these issues—we would expect an issue that combines all of these topics to be irresistible to journalists. Each separate issue called into question the legitimacy of various organizations and people, and this gained media attention. We would expect journalists to cover UCB-N given the history of past practices. What could not be predicted with any certainty in advance was where journalists would fall on a continuum with defenders of capitalism at one end and defenders of academic research untainted by corporate influence at the other.

The research agenda that often develops around a social problem consists of tracking the actors involved in the debate, situating the debate within the public arena where it appears, and studying the general slant of the discourse that surrounds it, including changes in evaluative positions. In addition, the research agenda may include analysis of public perceptions of the debate (e.g., see Gaskell and Bauer 2001 for a discussion of a research agenda concerning the biotechnology debate within policy arenas, media discourse, and public opinion in various countries). A typical methodological approach is to perform content analyses on public documents, such as news articles, to offer some kind of insight into how the issues are being, or have been, presented. Before turning to coverage in the popular press, let us look at UCB's public presentation of the Novartis agreement.

THE PUBLIC RELATIONS CAMPAIGN

It is often hard to discern whether an organization develops a public relations campaign so as to be proactive about a situation, anticipating problems that may arise down the road, or simply in reaction to public input, both positive and negative. The information about UCB-N that initially came from UCB seemed designed primarily to give the agreement a positive "spin." The timing and content of UCB press releases show that the university was then forced to react to criticism from actors off campus. Our reconstruction of UCB's PR campaign relies largely on articles that appeared in *The Berkeleyan*—the faculty and staff newspaper at UCB—and the website of the UCB Office of Public Affairs (known as the Public Information Office when the agreement was signed). We analyzed nine articles from *The Berkeleyan* and four press releases from the public affairs website. We found these thirteen documents using the Internet, though other articles and press releases may have appeared that were not available to us. Six of the articles and news releases appeared between November and December 1998, three appeared in 1999, two in 2000, and two in 2003. It should be noted that these two sources are not identical, though both are available on the Internet. Press releases were given to reporters and made available on the website, while *The Berkeleyan* was available to faculty and staff on campus. The Office of Public Affairs is in charge of both sources of information.

As mentioned earlier, the dean's office in CNR and leading faculty in PMB controlled the release of most of the official information concerning UCB-N prior to its signing. The first article to appear in the popular press was published in October 1998, while the first four press releases were made available on November 23, 1998, the day the agreement was signed.

The UCB press releases that appeared on that date emphasized the positive aspects of the agreement. One was composed of three quotations from representatives of California's agricultural industry, all of them from outside the university and all praising the potential impact UCB-N would have on the agricultural community.[1] Another, written in a question-and-answer format, asked: "Will NADI be able to influence internal department policies or direct the research agenda?" "No," was the answer.

> The two NADI members of the Research Committee will have no voice in department or University policy. The research agenda will continue to be determined by the interests of individual PMB faculty who may or may not choose to apply for NADI funding. Most faculty researchers have indicated they will participate, but virtually all will also continue to seek funding from federal and state agencies and foundations, and several will continue to work with other private companies.
>
> NADI and Novartis Agribusiness Biotech Research, Inc. (NABRI) members of the Advisory Committee can only make recommendations to the University and to the department, they cannot set policy. (University of California 1998)

The other two UCB statements released at this time provided background details on the agreement, again emphasizing the fact that it was a positive (and experimental) arrangement between UCB and Novartis. "This is the first, though experimental, step in what we hope will be a long and fruitful relationship," said UCB chancellor Robert M. Berdahl. "Novartis brings significant intellectual assets as well as financial support to an important area of fundamental research" (Mena and Sanders 1998). We have already discussed the belief, shared by a number of UCB people, that UCB-N would be an experiment, but this statement shows that this fact was also part of the public presentation and presumably was intended to allay fears of any long-term detrimental consequences.

The next mention of UCB-N came on December 2 in *The Berkeleyan* and reiterated that the agreement had been signed and that this "unique collaboration will keep Berkeley scientists and California farmers at the forefront of agricultural biotechnology" (Sanders 1998). This article mentioned the meeting between SRR and Rausser and implied that UCB administrators were taking positive steps to address internal opposition, and that students had the ear of the dean. The pie-throwing incident was depicted as the work of a group of "rogue ... anti-biotech activists" who had "sneaked into the press conference." They were "subdued by campus police" and "the press conference quickly resumed" (Sanders 1998). The article also suggested that the research to be accomplished under UCB-N was likely to yield rapid benefits for California agriculture.

The first article to include any critical commentary was published in *The Berkeleyan* on February 17, 1999, nearly three months after the signing, and focused on a talk at UCB given by David Noble of York University that was crit-ical of the commercial and industrial presence at universities. Noble critiqued everything from UCB-N to UCB's exclusive contract with Pepsi for soda pop vending machines. The article quoted Rausser, who suggested that Noble either had not read the agreement or had misunderstood it. Noble was not given the opportunity to respond, reinforcing Hoynes and Croteau's (1991) contention that controversy is often spun to suggest that criticism of internal policies comes only from uninformed outsiders who often harbor bad intentions.

The other two pieces of information coming from UCB in 1999 were gen-erated by the Office of Public Affairs and offered cautious optimism about UCB-N. In one of them, the chancellor said that, while the agreement would be closely scrutinized, such arrangements were going to become more com-mon in the future as public funding became less available for research, and that in this case the promise of genomic research was too important to ignore. In the other press release, a distinguished UCB professor argued that, while not everyone was happy with the terms of the agreement, it might serve as a catalyst for putting a more rigorous research protocol in place.

Only two UCB articles appeared in 2000, both in *The Berkeleyan*. One simply mentioned that the agreement would need to be studied in more detail as it unfolded, and the other used the agreement as an example of corporate linkages that would bring money to universities but risked loss of control over research agendas. This second article stated that the Berkeley Faculty Associ-ation had joined the American Association of University Professors, a group that was looking into concerns about both private and public funding at uni-versities. Nothing else appeared in *The Berkeleyan* until 2003.

On January 29, 2003, an article in the campus paper focused on the events surrounding UCB-N as it neared its conclusion (Sanders 2003). This article mentioned that Syngenta had not discussed extending the contract with the university, though an informal conversation between a public relations offi-cer and Syngenta researchers revealed that if the collaboration was to continue it would be much more limited in scope. The article presented supporting and opposing viewpoints, using such language as "positive attitudes toward the agreement," "students have been among the biggest beneficiaries of the agree-ment," "a contract with the devil," and "commercial interests." While this was the first article to quote individuals connected to UCB who were openly crit-ical of UCB-N, of the twenty-eight paragraphs, thirteen were neutral, ten pos-itive, and only five negative. The final *Berkeleyan* article, which appeared on September 25, 2003, mentioned UCB-N as one of a long list of accomplish-ments under Chancellor Berdahl's tenure at UCB.

These articles and press releases make clear that the generally accepted values and practices of the university—academic freedom, creativity, the use of outside funding sources—were, for the most part, being upheld. Some adversarial viewpoints were presented, though these were given minimal coverage. None of this should be too surprising, given the role of a public relations office. In fact, a caveat is in order here: These press releases and articles, while available to a large audience, did not necessarily generate a reaction, including from readers on the UCB campus. The point is that the Office of Public Affairs was presenting UCB-N in a positive way to anyone who would listen. We do not know whether readers found these articles and releases interesting or insightful, though we do know that this issue was also covered by newspapers and other mainstream media outlets off campus.

EARLY COVERAGE IN THE POPULAR PRESS

We collected articles from mainstream media outlets through a Lexis-Nexis search,[2] newspaper archives (*San Francisco Chronicle, San Francisco Examiner, San Francisco Bay Guardian, San Jose Mercury News, Contra Costa Times, Daily Californian*—the last the student newspaper of UCB), a general Internet search, and articles provided to us from people at UCB. We identified seventy-one articles published between October 1998 and June 2002, of which twenty-nine (40.8 percent) appeared between October and December 1998,[3] which demonstrates that coverage declined rapidly after the agreement was signed. This is unsurprising given that research performed under the agreement had not resulted in controversial technologies or products. It should be noted, however, that some news reports appeared in 2002 after Ignacio Chapela's work on genetically modified genes reportedly found in Mexican corn became public. Some of this coverage mentioned Chapela's opposition to UCB-N, as discussed earlier. We shall return to that incident in a later section.

Before turning to the articles, it should be noted that we, like most people involved in media analyses, were seeking patterns in seemingly chaotic environments. These patterns are typically referred to as "frames" and are viewed by reporters and audience as mere reflections of the natural order of things. "In short, news presents a packaged world, and not all of the recesses of the package are visible" (Gamson 1984, 80). Gamson and Modigliani (1989), for example, studied editorials and political cartoons as a means of exploring how editors think about issues, as well as changes in the coverage of nuclear power over time. In this case and others, the discovery of patterns in media representations—which are typically imperceptible when reading a newspaper on

TABLE 7.1   News Frames in the Reporting of UCB-N

Frames	1998	1999	2000	2001	2002	Total
Pandora's Box	8	8	11	5	6	38
Economics	11	1	0	6	1	19
Progress	7	0	1	1	0	9
Ethics	2	0	0	0	1	3
Runaway technology	1	0	0	0	1	2

a daily basis—illuminated processes internal to the workings of the media and the creation, maintenance, and challenges of social problems.

In their study of nuclear power, Gamson and Modigliani (1989) found that "progress" frames (nuclear power is a good thing) dominated early coverage, followed later by frames such as "Pandora's box" (only bad will come of this) and "runaway technology" (technology is out of control and it is too late to turn back). The latter two frames tended to coincide with the environmental movement of the 1970s and the accident at Three Mile Island in 1979. This sequence of frames was not followed in the coverage of UCB-N, as reporters tended to frame the issue in a negative way from the beginning, using the context of private funding and corporate control of higher education to set the stage. Reporters followed local practices as they sought ways to make sense of the issues in a way that would be accepted by colleagues and readers. This included taking into account such factors as the history of UCB as a campus known for student and faculty activism, the controversial nature of biotechnology, and declining government funds for research.

While other frames were used throughout the coverage, one of our more interesting findings is that the "progress" and "Pandora's box" frames were present from the start. Table 7.1 shows the number of stories presented within these frames and others between 1998 and 2002. While the Pandora's box frame was introduced at the outset and dominated throughout the coverage (accounting for thirty-eight of the seventy-one articles), the second-most-used frame was economic (money/financial benefits and/or risks), present in eleven of the 1998 articles. Of the nine articles coded as having a progress frame, seven appeared in 1998. A more general ethics frame—a framing of the agreement as positive or negative for the university without drawing on the other frames (such as supporting progressive science, the economics of university funding, and so forth)—was used twice in 1998 and once in 2002, and one runaway technology article appeared in 1998 and another in 2002.

In short, while Pandora's box was the most consistent and dominant frame relative to the others, it did not completely subsume them. Still, those who

TABLE 7.2   Characteristics of News Coverage

	Count		Count
Theme		Section	
Corporate control	57	National news	24
General research	48	Local news	13
Economic concerns	46	Commentary/op-ed	11
General ethics	16	Business	11
Food/agriculture	12	Science	2
Education of students	7	Other	10
Actors (top five)		Word count	
UCB administrators	39	Less than 500	31
Non-UCB faculty	22	501–1,000	28
UCB, not PMB, faculty	19	1,001+	12
Activists	18	Author	
UCB students	16		
		Local reporter	41
Page		Wire story	13
Front page/front of section	12	Outside commentary	11
Other	59	Other	6

followed this story in the press were often led to believe that only bad things would come from UCB-N—mainly in the form of private business control over campus research. For example, an Associated Press story covering the California state senate meeting on the contract stated, "Senator Steve Peace, D–El Cajon, blasted the UC system for what he called 'inherent conflicts of interest' in which private companies help decide what research is funded and who gets the money, often without the public being any the wiser" (Thompson 2000).

Framing is a conceptual tool used for media analysis that is open to numerous interpretations, and while we think it is helpful, it is also important to think more deeply about the context in a way that is not as abstract or susceptible to idiosyncratic interpretations. We categorized various themes in the news articles and found that corporate control of higher education was the dominant theme, appearing in fifty-seven of the seventy-one articles. The second-most-discussed theme was general research, appearing in forty-eight articles, followed by economic concerns (46 articles) and general ethical issues (16 articles). Activism was a popular theme in 1998 (6 articles), as reporters covered the pie-throwing incident mentioned earlier. However, this theme appeared in only one article in 1999 and three in 2000. Coverage of corporate control over higher education continued to dominate coverage throughout the five-year period. Table 7.2 summarizes the various aspects of the media coverage.

The change in coverage from specific events—e.g., the pie throwing—to more general issues about corporate control and (perceived) economic necessities within universities tracks the evolution of this story. The metaphors gradually disappeared, as did the activists and student protests, but the general topic remained important to some degree.

With regard to the people covered in the news, UCB administrators were the most prominent, appearing in thirty-nine articles. In addition, activists appeared in eighteen stories, UCB faculty in twenty-five, UCB students in sixteen, and Novartis representatives in twelve. Faculty and administrators from other universities appeared in thirty articles, emphasizing the degree to which UCB-N was an issue at other institutions of higher education—a subject discussed below.

If these articles had appeared on the front page of newspapers, this would be evidence that the public and editors were deeply concerned about the UCB-N story. In fact, however, only four of the articles appeared on the front page (two in the *San Francisco Chronicle,* one in the *San Diego Tribune,* and one on the cover of the *Atlantic Monthly*), and another eight were on the front page of various sections of different newspapers. This means that forty-five of the articles appeared on the inside pages of newspapers and other publications, including fifteen from wire services that could have been used by any number of newspapers as filler material or been completely ignored. It should be emphasized that some of these articles did not come from mainstream newspapers (nineteen appeared in smaller sources outside the mainstream, though the article in the *Atlantic Monthly* was the major cover story). Twenty-four of the articles appeared in national news sections, thirteen in local news sections, and eleven each in the opinion pages and business sections. Only two appeared in science sections. Most of the articles were written by in-house reporters (41), followed by wire stories (13). Forty of the articles included some kind of negative commentary by the author, twenty-two had no commentary, and nine contained some positive commentary by the author.

The headlines of these articles are also important. Headlines give a specific slant to a story, so that even a report that praised UCB-N may be read as derisive if accompanied by a negative headline—not an unusual situation, as headlines are not written by the reporters themselves (e.g., Parenti 1993, Pfau 1995). Of the seventy articles with headlines, thirty-five of the headlines were negative, twenty-seven were neutral, and eight were positive (one article was a wire service story with no discernable headline). If it is true that readers encounter headlines on a random basis, they were much more likely to come across a negative headline than a positive one. Given the frequency of negative frames and the prominence of activist opponents of UCB-N in the stories, much of the coverage offered by the popular press was less than favorable. Sometimes the negative slant appeared in the form of emphasis on the statements and activities of UCB-N opponents, while at other times the articles seemed interested in stressing the negative aspects of the agreement.

POSSIBLE INTERPRETATIONS

While counting and coding sources and themes is relatively straightforward, trying to measure public reaction is not so easy. For example, in a study of political party members prior to a major election, Gunther (1992) found that Republicans tended to think the media favored Democrats, while Democrats tended to think the media favored Republicans. While we coded our overall evaluations of the articles—forty-six were either very or somewhat critical of the agreement, twelve provided both positive and negative aspects, and thirteen were either very or somewhat positive—we felt that a more active approach was needed to serve as a proxy for reader engagement.

One way to approach such an undertaking is to think of the issue as a breach of a norm. Schonbach (1990) studied how people react when faced with a situation in which a social norm has been broken and found that when people were asked to account for their actions, if they refused to give an account or provide a justification (putting the blame on an external force/actor), this tended to escalate conflict between the parties. When they made excuses (putting the blame on an internal force) or offered a concession, this tended to manage or defuse conflict. Accounts were given in forty-one articles, of which thirty-eight were justifications, two were refusals, and one was an excuse, and no concessions were reported. The most popular justification given was by UCB administrators and faculty who said that the Novartis money was needed because public research funds were drying up. According to Schonbach, this strategy would lead to a higher level of conflict between the opposing sides. It should be noted that supporters of UCB-N may have viewed the situation as normal rather than as a breach of the normative structure of university activities. This may have had an effect on what type of account they gave, as people who felt they were in the right tended to ignore or belittle their opponents, which may have added fuel to the fire.

One interesting note is that the only excuse given appeared in the popular press, while one of the refusals appeared in the popular press and another in the UCB student newspaper. In other words, all audiences (of the 71 articles coded, 51 appeared in the popular press, 8 in the UCB campus newspaper, 4 in scientific journals, 2 in higher education journals, and 6 in other sources) were receiving the same accounts—justifications. Actors involved in the coverage were not changing their stories as the issue moved to various media outlets, nor were they changing their stances as the story evolved among reporters and media organizations.

If Schonbach is correct that refusals and justifications lead to greater conflict, we would assume that people interested in this topic and reading these reports would have found the accounts of UCB administrators unacceptable. For example, in an Associated Press story released on May 15, 2000, a state official was quoted as saying, "California universities have little choice but to seek private research money because state funding has failed to keep up with rising costs" (Thompson 2000). For opponents, this kind of statement obscured the fact that the university seemed unwilling to admit that it could seek and distribute funds in other ways, and that private funding may come with strings attached. This kind of reaction did dominate the news during the early days of UCB-N, for example in an editorial in the *San Francisco Chronicle* charging that the contract had been negotiated behind closed doors. "We … cannot ignore the implications for academic freedom in a college beholden to corporate interests" (Rosset and Moore 1998). Part of the concern, as discussed earlier, was that the funding would go to a whole department instead of a single individual or lab. This quotation came from an editorial, and it should be noted that while editorials—especially those written by individuals with vested interest in the topic— and regular news items may be interpreted differently by readers, both types of information can be found in the same place, making it difficult to determine their exact impact on public opinion. It is not hard to see this editorial as an attempt by individuals with agricultural interests to escalate the conflict.

REACTIONS TO THE EARLY COVERAGE

The actual news coverage is only part of the story. We must also look at how immediately interested readers reacted, demonstrating how the coverage affected those with vested interests in the agreement, including people with only indirect ties to the contract. An article that appeared in the *Chronicle of Higher Education* shortly after the agreement was signed noted that it was "sparking fierce debate on the Berkeley campus—and throughout academe— about the influence of industry money on academic research" (Blumenstyk 1998, A56). This debate within the academic community did not subside. For example, a letter to the editor of the same publication, published in 2001, contended that, while an earlier article in the *Chronicle* had denied that opponents' worst fears had been realized, there was no way to prove this. "Demanding that critics provide instances of overtly compromised research encourages non-overtly compromised research" (Arzoomanian 2001).

This shows that while the effects of UCB-N were mixed, some individuals within higher education were unwilling to travel the same road. The ways in which the media framed the story had an impact on how readers interpreted it. In one article a government official not affiliated with UCB was quoted as saying that UCB-N was having a negative impact on the government. Even though UCB-N concerned UCB and a private company, this official thought that readers would jump to the conclusion that the government was involved, presumably because other biotechnology research has often been funded by government agencies.

We gathered responses from people beyond the UCB campus for the most part through secondary sources. In addition, we conducted interviews with people on campus, as well as a few others from the Bay Area, which shed some light on how the public debate was affecting them. As noted in earlier chapters, most UCB individuals we interviewed were quick to condemn negative media coverage of UCB-N, though most of these people approved of the agreement, so their views of the media coverage are what one would expect. As one UCB administrator commented, "the agreement was expensive to UC in terms of negative public relations."

Others, however, benefited from the coverage, as it gave them a way to use the history and mythology of Berkeley to make the story interesting. A Bay Area journalist, for example, said that UCB-N had parallels to the free speech movement of the 1960s, which gave it an anchor with deep resonance both within and outside Berkeley regardless of the actual relevance of the comparison. The UCB campus was supposed to be democratic, and administrative decisions were supposed to be made with a high degree of transparency. This was the rationale on which the movements of the 1960s relied in part, and some saw the UCB-Novartis dealings as a return to negotiations behind closed doors.

One administrative specialist argued that "we would be worried if there wasn't any press, because this agreement may signal a shift in how university-industry relations are constructed," while another UCB administrator said that journalists had blown the whole thing out of proportion, but that this was not surprising given that journalists were only trying to sell their stories to consumers. Both sides agreed, however, that UCB should have been involved in the public relations campaign, as both journalists and consumers were not knowledgeable enough to interpret the story on their own. These are standard third-person responses (Davidson 1983; Ten Eyck 2000), in which others, including journalists, are considered to be more deeply influenced by media coverage than oneself. As one professor put it succinctly, the coverage was "inflammatory," though one must assume that this individual was a supporter of UCB-N.

It should be noted that while the majority of comments about press coverage from UCB employees were negative, some saw it as either balanced or in some way a positive contribution. One professor referred to an article in the *Atlantic Monthly*[4] as "spot on" in discussing the relationship between universities and industry, while a UC dean said that the range and degree of all the publicity "may lead to future contracts." These various readings of the reporting show that issues reported in the media are rarely treated as homogenous across media sources (Gamson et al. 1992), and that interpretations of the same information will vary from person to person (see also Hoijer 1992; Morley 1980) and may even change for the same person as they move from one social context to another (Fiske 1992).

THE AFTERMATH

The controversy over UCB-N did not end when the money stopped flowing. Twelve articles appeared in the Lexis-Nexis archives between January 2003 and August 2005 (all in western news sources), two of them on the front page. All twelve articles continued to question the agreement, with much of the focus on the denial of tenure for Chapela. The *Los Angeles Times* summed up much of the coverage:

Anyone who has witnessed the aftermath of a fire knows that the more acrid the smoke, the longer the stench persists. That could be bad news for UC Berkeley, which was plainly hoping to dissipate a foul cloud last May when it finally awarded tenure to Ignacio Chapela as an associate professor of microbial ecology.

With the decision, Chancellor Robert Birgeneau reversed a 2003 ruling by his predecessor, Robert Berdahl. But it's doubtful that his move will entirely dispel the ill will that presumably led to Berdahl's decision in the first place.

At the heart of the case was a five-year, $25-million agreement between Berkeley's department of plant and microbial biology and the biotech company Novartis. The agreement expired in 2003, but it still marks the dangerous path that universities must tread as they turn to industry to replace government funding, which has been drying up or arriving with political strings attached.

The Novartis agreement, as a team from Michigan State University observed in a report Berkeley commissioned, "highlighted the crisis-ridden state of contemporary public higher education in California ... and the country." (Hiltzik 2005)

The main point of this article was that university-industry contracts were difficult to navigate, and that UCB-N was likely to leave a pall over the Berkeley campus because of its links to the denial of tenure to a popular professor. Other articles discussed concerns with biotechnology, academic freedom, and the danger that industry agendas could subsume academic ones. The coverage has not inundated the general public, but it does seem to be having an effect on many within the ivy-covered walls of American academies.

The media practice of putting a spotlight on the downfall of others clearly seems to have been at work in this story. At the same time, while the role of the media in bringing the agreement between UCB and Novartis to the public's attention cannot be overlooked, it should not be overstated. As Dearing and Rogers (1996) contend, the media present issues for public consumption, but they do not necessarily tell us how to think about them. Individuals at Berkeley, in California, and in the United States are immersed in information. When UCB-N was signed and made public, candidates and voters were considering a midterm election (November 1998), the Clinton-Lewinsky affair was still fresh in many people's minds, fighting continued in the former Yugoslavia, many of the Asian Tigers were in a financial tailspin, and a California baseball team (the San Diego Padres) was in the World Series. All of these stories, and many others, were vying for attention. Some people were inclined to be interested in the story, while others were not.

The story of UCB-N did have real and perceived impacts on the UCB campus and beyond, however. The issue hardened existing divisions between faculty and administration, between different colleges, between the university and government, between industry and the university, and between private citizens and UCB faculty, and may have created some new splits as well. Many observers thought reporters were focusing on the negative and sensationalizing the contract, causing some UCB personnel to think about how PR campaigns could be conducted differently in the future. The fallout from news coverage is often overlooked by media scholars, but it is perhaps the pivotal measure of the effectiveness of the press. Whether or not similar agreements between universities and industry will be handled in the same way in the future or given the same amount of attention is an empirical question. It is clear that, at least for the time being, media coverage has changed both individual and collective behavior on the campus of UCB regarding money coming from industry.

EIGHT

The Scientific Enterprise

MUCH OF THE CONTROVERSY generated by UCB-N stemmed from the potential consequences such an agreement might have for the various constituents of PMB. There were questions regarding the effect of UCB-N on the research directions of faculty, postdoctoral researchers, and graduate students: Would the large amount of noncompetitive money from UCB-N lead faculty to pursue and secure fewer competitive grants? What might the effects be for teaching, at both the undergraduate and graduate levels? Many of the questions about the agreement that were posed by DIVCO and other concerned parties to Rausser, PMB, and the administration could not be answered in advance and resulted in the decision to treat UCB-N as an experiment.[1]

In this chapter we present views of UCB-N from the multiple standpoints of PMB members—graduate students, postdoctoral researchers, staff, tenured and untenured faculty. We also examine the material and cultural consequences of the agreement for PMB, to the extent that we were able to discern them, in order to answer some of DIVCO's questions. This section ends with a discussion of a number of issues raised by PMB's close department-wide involvement with a private research corporation.

VIEWS ON THE PARTNER SELECTION PROCESS

There was much discussion among the PMB faculty with respect to the desirability and shape of any agreement with industry prior to the visits by the C4 to companies that had expressed interest in a potentially department-wide agreement. Those most closely involved in establishing UCB-N said it took a while to get faculty to see that it was okay to reverse the usual form of agreements and in effect have companies compete for the honor of allying with PMB. Many faculty members commented on the "endless" discussions on the ethics of such an agreement and whether they should go with just one company. In the end the department concluded that the single company route would preserve the department's academic freedom, whereas an agreement with multiple companies would be more likely to balkanize PMB. They also decided that IPR would be problematic if several companies were involved. All but two PMB faculty members agreed with the "one company" strategy.

Wilhelm Gruissem, former PMB professor and the initial PI for UCB-N, described the solicitation process as open and transparent until the proposals were received from Novartis, DuPont, Monsanto, and Pioneer. At this point the companies were told that the details of their proposals would not be disclosed to anyone other than the PMB faculty and relevant senior administrators. As noted earlier, anyone who wished to see these proposals had to sign a confidentiality agreement confirming that they would discuss the contents of the proposals only with those who had also agreed to confidentiality.

JUSTIFICATIONS FOR ENTERING
THE AGREEMENT

Three different justifications were offered to explain why PMB sought an agreement with an industrial collaborator. Within PMB, only faculty members were able to provide explanations as to why their department sought a university-industry agreement of the design that resulted in UCB-N. The predominant argument was that scientific research is expensive and that academic units do not have the resources (financial or material) to keep up with private industry in this respect. Thus departments such as PMB require industry funding if they are to remain on the cutting edge. Biology, for example, particularly in its molecular and genetic disciplines, has moved into the realm of "big science" with the use of very expensive and/or proprietary technologies; information is necessary to move biological knowledge forward by even a tiny step. Russell Jones, a PMB professor, said, "What we did rely on was [Novartis's] technology. But you know, this is how science is today; that you cannot do it with a piece of string and a piece of chewing gum anymore. You have to have this highly technical infrastructure and this is what we got out of TMRI."

The second justification concerned the benefits that were expected to accrue to PMB from its recovery of indirect costs. PMB had significant departmental debt, no management service officer, only one-fifth of a secretary's time, and no prospect that the departmental budget from UCB would be enough to cover operating costs anytime soon. Again, Russell Jones provides the clearest example of this justification: "We pay for our own telephones, we pay for our own computers, we pay for our own office furniture, so there's nothing that the university goes out of its way to provide. And you know, I can't blame the university, but if a department wants to run a program that's a cut above, then the department has to find alternative sources of funding. And research grants to individuals don't do this, because if I have a research grant, it's for me." It was expected that through PMB's recovery of indirect costs, UCB-N would support the daily operation of the department and thus make it easier for an individual to conduct research.

The third justification centered on UCB-N's benefits for PMB's graduate students. Even though there was general encouragement from the chancellor and president to enter into closer relations with industry, PMB faculty said that UCB-N was initiated by the department on the basis of PMB's assessment of its own need to establish a good graduate program. Indeed, Wilhelm Gruissem asserted that the original goal of UCB-N was to support the graduate student program.

REASONS FOR NOVARTIS'S INTEREST

Plant and Microbial Biology Faculty Perceptions

Apparently discussions were never held among PMB and NADI scientists on the specific reasons why the company wanted to enter into a collaborative research agreement with an academic department, but this did not prevent PMB members from speculating on the possible motivations. Several people thought that UCB-N was a way for Novartis to explore the new field of agricultural biotechnology cheaply at a time when the direction of the entire company was being reconsidered. The agreement gave the company access to much of the research done by the entire department, including postdocs and graduate students, without having to pay their salaries or benefits. Indeed, if Novartis had tried to replicate the profile of the department at NADI, a member of the PMB faculty argued, "it would be a lot more than $5 million a year, for five years, just to pay all those salaries alone."

When asked to speculate as to why Novartis wanted to enter UCB-N, another faculty member replied, "because I think, if you don't mind my saying so, we have really excellent plant scientists in this department and I think they [Novartis] really wanted to enrich their own research program." While

Patricia Zambryski, another PMB professor, thought that Novartis was more interested in having the direct scientific exchange with PMB scientists than in focusing solely on IP, Sydney Kustu argued, "it is my sense that Steve [Briggs, NADI CEO] is also bright enough to realize that the biggest things to come out of the alliance in terms of research are things that he did not foresee. So he foresaw enough to say, 'Hmm, we will get adequate return,' and he knew enough to know that he probably did not foresee some of the things that might turn up."

Novartis/Syngenta Explanations

By and large the explanations offered by employees at Novartis and Syngenta for their involvement in UCB-N do not differ dramatically from the speculations of PMB faculty. The merger of Ciba-Geigy and Sandoz that created Novartis really "inhibited the organization from taking new initiatives," according to one employee. One result was that competitors such as Pioneer, Monsanto, and DuPont began exploring genomics much earlier than Novartis did. In 1998 Novartis realized that it had fallen behind and needed some way not only to catch up but to surpass its competitors. Thus the agreement with PMB was a way "to accelerate the rate of discovery, the basis of agricultural traits, beyond what we could do with our limited staff and facilities at that time." Novartis's plan was to build in-house capabilities as well as make use of academic expertise; while the former would take several years to establish, the latter could be used to jumpstart the program.

Two employees involved with the agreement maintained that it was established with PMB rather than any other group of public scientists because PMB offered a very broad range of academic interests (disease, development, transformation, plant biology) and is populated by very clever people. They said that the company did not really know what it wanted from such an arrangement, which was why UCB-N disbursed unrestricted money. If Novartis had known precisely what it wanted, then the company would have entered the usual restricted relationship with one faculty member or laboratory to do work in a particular area. While securing new IP was important, in that the company always sought to increase its patent portfolio, it was not the primary motivation for entering the agreement. Instead, Simon Bright, head of technology interaction for Syngenta, described the company's mindset as one of "let's be surprised" by what PMB can do. In a similar vein, Bright declared that Syngenta would not have been interested in setting up an agreement similar to UCB-N elsewhere because the other places did not have the same confluence of intelligence and ability.

Uniqueness of Novartis Agricultural Discovery Institute

Prevalent among many people on campus was the notion that NADI was unique and thus that UCB-N was an unrepeatable alliance between two specific entities at a particular moment in time. In large part this notion stemmed from the structural characteristics of NADI: It was part of the research-oriented Novartis Foundation rather than the product-oriented Novartis Corporation. Thus NADI was buffered to a degree from the business concerns of the corporation. Peter Quail, PMB professor and research director of the Plant Gene Expression Center, said that he thought NADI's initial attractiveness was partly due to Steve Briggs's knowledge of and desire for an academic environment that was in turn enabled by NADI's research independence, accompanied by monetary support from Novartis. Scott Kroken, a former PMB student, postdoctoral researcher, and TMRI employee, concurred, suggesting that "the agreement was really quite natural between this company [NADI] and the university because this company really functioned in a sort of quasi-university manner in the first place." In describing those she collaborated with at NADI, Talila Golan, a PMB postdoctoral researcher, said, "the people I interacted with were hired as scientists, not as industry people; they were hired to think and take a project and run with it wherever it goes, so they had all the thinking of an academic scientist rather than industry people."

As noted earlier, Briggs's background and already established personal relationships with people in PMB was another factor deemed important to the choice of NADI as PMB's partner. Briggs's experience as an academic was an especially significant point that distinguished him from other industrial scientists. In addition, while at Pioneer, Briggs had collaborated with academics at the Cold Spring Harbor Laboratory, whose plant biology program was headed by Nobel Prize winner Barbara McClintock. The program was essentially defunct, but Briggs restarted it with money from Pioneer and it is functioning today. Briggs also started two programs at Pioneer to permit academics to have access to Pioneer's plant gene sequences and functional genomics tools. Such a history and apparent conviction to work with academics sets Briggs apart from most industry scientists. As one TMRI employee put it, "most companies have no interest in providing access to their data or technology. You look outside of agriculture and it just doesn't happen, full stop. In agriculture, I think you could trace the instances of it back to initiatives that [Briggs] made both at Pioneer and at Novartis, and that's put pressure on the competition to do a little bit, but not much. I mean, still, the only organizations today that really work collaboratively are Pioneer and Novartis. And of

Novartis, it's basically TMRI. Providing this access is just not considered desirable and there's very, very few companies in the world that will do it."

With the dissolution of TMRI in 2003, however, the broad, department-wide type of university-industry collaboration exemplified by UCB-N appears to have disappeared. One former TMRI employee said, "the main reason I think TMRI was closed down by Syngenta, its new parent company, was because the research was really too empirical, of the type where the products were just too far down the pipeline." Because TMRI was not separated from Syngenta in the same way that NADI was from Novartis, TMRI had no opportunity to be a quasi-academic research laboratory. While some people thought TMRI was terminated and UCB-N not renewed because of the change in management, others said that the biggest change in the move from NADI to TMRI was in the market—that is, because the economy had weakened, it became more important for TMRI to create products quickly. In addition, Bright said that TMRI was not as "blue sky" as the PMB people seemed to think it was—it had always been geared to the creation of commercial products.

VIEWS OF THE NEGOTIATION PROCESS

Faculty Views

On the whole, the PMB faculty viewed the negotiation process as uncontroversial and straightforward. Faculty members were generally satisfied with the extent and degree of their involvement in the negotiations and trusted their representatives to create an agreement that would benefit PMB. One associate professor said that before the decision was made to go with Novartis, the departmental meetings focused on whether it was a good approach to take; after the decision, the meetings became more informational, as only a few people were involved in the actual negotiations. Not all faculty members participated to the same degree. Sydney Kustu, PMB professor, described being "carried along by those who wished, initially at least, to forge the agreement for plant science." Even so, another professor said that one of the things that surprised him was that the meetings were always well attended by faculty, half of whom often did not attend regular faculty meetings. He further described the PMB meetings as a spirited and openly democratic sharing of ideas, which suggested to him that the faculty were very interested in getting a good contract with private industry and wanted to shape the final terms of agreement. Most faculty members agreed that the departmental discussions about UCB-N were geared toward building consensus, though several professors were concerned that discussions centered on whether PMB was being sold to Novartis rather than on the agreement's potential consequences, positive or negative, for graduate students and the teaching mission.

By all accounts, it was in the months immediately surrounding the decision to go with Novartis and during the negotiations that PMB, as a department, grappled with questions about the appropriateness of an entire department at a public university entering a research agreement with a private corporation. Similar questions related to the appropriateness of individual faculty contracts with industry appear not to have been considered previously. Therefore, one early consequence of UCB-N was that it created a space for important discussions about the character of university research. Regrettably, this space seems to have been limited to PMB and only existed during the early phase of negotiations.

Graduate Student Views

By contrast to the faculty's view of open and democratic negotiations, many PMB graduate students felt excluded and deliberately kept in the dark about an agreement that was being proposed partly for their putative benefit. This was not a new complaint, for the general lack of communication and involvement of graduate students in departmental affairs had been a point of contention even before UCB-N. According to most of the PMB members, no graduate student or postdoc representatives attended the faculty meetings where UCB-N was discussed. Indeed, graduate student representatives had been excluded from all faculty meetings held since September 9, 1998 (PMB Graduate Students 1998). Scott Kroken, then a PMB graduate student and now a bioinformatics staff scientist for the Diversa Corporation, explained that Gruissem handpicked Susan Jenkins, a student in his lab, as the graduate representative and that she was subsequently held up as an example of graduate student involvement. However, because Jenkins was a member of Gruissem's lab and he was the faculty member most deeply involved in creating UCB-N, Kroken said that the rest of the graduate students did not consider her capable of representing them. Jenkins, in turn, was not at liberty to tell the other graduate students what went on in the meetings. Thus, while she acted as a token presence, she was not a true student representative.

While several graduate students and postdoctoral researchers paid little attention to the negotiation process, others were angry about their exclusion. One PMB graduate student said that the agreement was "touted as being done for grad students when they didn't know what happened." Another observed,

"All graduate students really resented being left out of the loop. We were given less information than available to the media; we were just expected to support it no questions asked. ... Faculty rammed the agreement through without consulting the graduate students" (quoted in MacLachlan 2000, 3).

As noted earlier, during the first half of 2000, Dr. Anne MacLachlan and Mary Crabb of the Center for Studies in Higher Education (CSHE) conducted a survey of PMB graduate students regarding their educational experience in PMB and their thoughts about UCB-N. MacLachlan concluded from the survey results that there were four main reasons why graduate students were unhappy with the process of negotiating UCB-N. These were (1) the secrecy of the negotiations in the department; (2) the fact that information was available to others in other departments and the press; (3) lack of consultation with the students; and (4) the kind of communication from faculty to students about the agreement before it was finalized (MacLachlan 2000, 3). Our study supports these conclusions and indicates that while there were initial attempts to improve collegial communication among PMB faculty and students, these efforts were largely transitory.

One graduate student who stayed on as a postdoctoral researcher said that graduate students expressed their opinions about UCB-N and its process of negotiation in individual conversations with Gruissem as chair of the department. Gruissem also came and talked with graduate students at one of their group meetings, but only after the agreement had been publicized. In December 1998 the PMB graduate students collectively wrote a letter to the PMB faculty that one student characterized as very controversial and confrontational and that did indeed create conflict between students and faculty. The letter laid out the main concerns regarding the exclusion of all but faculty members from the negotiations:

> One consequence of the restricted flow of official information is that graduate students have been forced to rely on rumors and supposition throughout the negotiation process with Novartis. Thus, we must take issue with recent newspaper quotes which claim that the agreement with Novartis was "arrived at through an open process" (Vice Chancellor Cerny's Letter to the Editor of the *San Francisco Chronicle*, 11/7/98). We are not aware of any meetings within PMB which provided information to or sought input from graduate students or post-docs. Apparently several open meetings were held within the College of Natural Resources, but most PMB students were not aware of these or were told that the meetings were open only to faculty. We find it puzzling that students in ESPM were informed of the CNR meetings while we were not. While

many students hope that the faculty simply did not realize the extent of graduate student interest in the development of the agreement, others find it difficult to believe that the consistent lack of information from official channels has been merely an oversight. (PMB Graduate Students 1998)

One positive consequence of this experience, at least in the short term, was that graduate students became more active in attending committee meetings and writing up their notes for everyone, and were generally more interested in departmental and university life.

Postdoctoral Researcher Views

The postdoctoral researchers in PMB at the time of the UCB-N negotiations also felt that they should have been included in the information stream. Paul Bethke, assistant research specialist, thought the approach taken by the department was unnecessary. He said:

> You were presented with this thing—"okay, now here's some stipulations that apply to you, which you have no say in, essentially." It was different for people hired later, where you could essentially say, "I understand the rules and I'm willing to accept those." It's a different situation to say, "I'm not willing to accept those and I'll just leave." My own feeling is that the faculty at least, and the college, did a pretty poor job of keeping people informed and knowing what was happening and why it was happening ahead of time. They just presented this agreement with all its stipulations as a done deal and people jumped on the stipulations and didn't spend any time thinking, Did it make any sense to do this sort of agreement?

Jennifer Vorih, former PMB student affairs officer, agreed with Bethke's assessment. She said that while graduate students had been more involved with departmental affairs before UCB-N, this was not the case for postdocs, who, she thought, "became a little more politicized because of this." If so, their politicization appears to have been short-lived. When Todd Leister was hired by PMB as a postdoctoral researcher in 2000, he heard very little about UCB-N and was given no information about the structure of the agreement. Indignation may have faded quickly because of the relatively high turnover of postdocs and their lack of involvement in departmental affairs beyond their specific lab affiliation.

IMPLEMENTATION OF THE AGREEMENT

Arguably, the collaborative research agreement between UCB and NADI has never been fully implemented, thus automatically curtailing the extent of its impact. In particular, a joint UCB-NADI research facility (appendix E of the contract) and the appointment of NADI employees to UCB academic positions (article 7.2 of appendix D of the contract) failed to materialize. We can therefore speak only to the consequences of the parts of the agreement that were actually put into practice.

With respect to the formal language and structure of the agreement, most of the people directly involved said that UCB-N was implemented in the way they expected. One faculty member said, "It worked out just painlessly." While some people were against the idea of UCB-N in principle, those funded through the agreement generally concurred with Zacheus Cande, jointly appointed MCB and PMB professor, who said that the implementation of UCB-N and its "practices have been amazing benign, I think, from my perspective." Others pointed to the small number of amendments added during the execution of UCB-N as evidence of the soundness of the original contract. William Hoskins, director of OTL, said there were "no major problems with the agreement that we couldn't handle with an amendment ... and these were not significant amendments." To a large degree this view is also held by those in charge of the agreement on behalf of Novartis, one of whom said that he does not think that anything was overlooked in UCB-N that would need to be changed to make the agreement more successful for his company.

There were also ways in which the agreement diverged in practice from what was expected, sometimes positively, sometimes negatively. Bob Buchanan, PMB professor, said that, "from the scientific standpoint, it exceeded expectations in the yield, in what's come out of it." At least one person was disappointed that Briggs's vision of UCB-N did not reach fruition. Sydney Kustu, PMB professor, lamented that Briggs's ideal of "a new and wonderful way in which people could cooperate" was not realized in the actual operation of UCB-N.

Collaborations

UCB-N also diverged from faculty members' expectations in collaborations with the NADI scientists. Several faculty members commented on the infrequency of communication between PMB and NADI personnel and the different levels of access to certain technologies both within and among laboratories.

These perceived drawbacks were closely tied to the failure to build an interface research facility that would have housed the NADI scientists, technologies, and information. In the words of Paul Bethke, assistant research specialist, "Originally, I think, people thought there would be in-house [at UCB] secure computers set up where we could access things on our own and it turned out ... that we were generally going through intermediaries, saying, 'We need this kind of information, can you send us something?' There would be a pause and then it would come back, but that again is a slow way of doing it."

Traveling to San Diego might have accelerated the pace of retrieval of information but would have increased the cost of the collaboration and disrupted other scheduled activities. The point is that a number of PMB researchers thought they would only have to walk down the hall to access the benefits of collaboration. Since the interface facility was not built, the "seamless interaction" had not worked out in the way some people had hoped.

Other faculty members said their collaboration with NADI scientists exceeded their expectations. Professor Russell Jones established an individual agreement with TMRI outside the purview of UCB-N, something he hadn't expected to do when he signed UCB-N, but "this is what has made this [UCB-N] agreement for me exceptionally beneficial. In that it's drawn me to these people simply because I was forced to be drawn to them. I'm not sure that I would have got involved with the likely people at TMRI if the agreement wasn't in existence."

Confusion

There were two main points of confusion regarding the implementation of UCB-N. The first was whether graduate students had to sign the general agreement or whether they would be covered automatically under their advisor's signature. This issue was skirted by changing the status of the UCB-N money used for graduate stipends to "19900" state graduate student researcher funds disbursed through CNR. This move obviated the need for students to sign confidentiality forms because officially they were not getting any money from UCB-N, nor were they doing research under its auspices. This situation changed after their third year in the program, when a student moved to being financially supported by his or her advisor, at which point it was decided on a case-by-case basis whether students had to sign confidentiality agreements.

The second point of confusion surrounded the requirement that all papers for presentation or publication were to be sent to NADI for a thirty-day advance review. Even two years into the agreement some people claimed ignorance of this obligation. As late as mid-2000 we were told that a faculty member said of a presentation, "I had no idea we were supposed to submit this." Richard Malkin, PMB professor and former CNR interim dean, confirmed this. "There was a meeting," he recalled, "where this was brought up, and the faculty were very upset. 'You mean we're supposed to send something to Novartis before we publish? Oh, I'm not going to do that.' And I thought, 'These are the people who just signed this agreement!'" Either these faculty members did not read the contract carefully before signing, or they chose to ignore or forget the parts they didn't like. Faculty who did submit their papers generally saw this as an inconsequential requirement. Several people simply sent in drafts or outlines and considered the thirty-day NADI review similar to other delays involved in the publication process. Indeed, some people submitted their manuscripts to NADI and a professional journal simultaneously. One positive consequence of this requirement was that some people wrote papers further ahead of time than they would otherwise have done.

Funding of Projects

The composition of the research committee established to evaluate the faculty proposals and disburse the UCB-N funds concerned many people because it had two representatives from Novartis in addition to the three PMB faculty members. According to those most closely involved in creating the structure of UCB-N, PMB invited the two Novartis representatives to serve on the committee; Novartis did not demand this. Bob Buchanan of PMB said the Novartis people were invited in order to get their insight and enable cross-fertilization between industry and university researchers; indeed, such an arrangement was not unusual. The research committee worked as most of the participants thought it would. The five members worked to reach a consensus based on three criteria about whether to fund the proposals and, if so, at what level. As listed in the research agreement (appendix B, 30) these criteria were (1) the quality and intellectual merit of the proposed research; (2) the potential advancement of discovery; (3) the past and present productivity of the PI. The three PMB members recused themselves from decisions about funding their own proposals. The committee members all received the proposals before meeting, and each member rated proposals according to the criteria listed above. These scores were sent to the administrative assistant, who entered them all into a spreadsheet; the committee then met to decide how to reconcile their disparate scores. The two NADI members participated by e-mailing their scores, through telephone conferencing, and on one occasion by a visit to UCB.

No faculty proposal was ever rejected, to the knowledge of one committee member, although there were different levels of support, ranging from

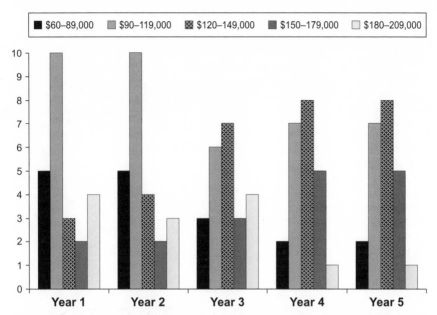

FIGURE 8.1 Number of UCB-N grants awarded each year by funding amount. The number of grants is not constant because two faculty members left and one joined over the course of the agreement.
Source: Source: PMB data.

$60,000 to $200,000 each year. Over the course of the five years, the total amount of UCB-N money spent directly on faculty research was $14,240,000, or 59 percent of the total amount of the agreement.[2] The average annual award was $120,500. Twenty-five PMB faculty members received money from UCB-N, but two faculty members left PMB during the agreement and therefore did not get funded for all five years. PMB hired another faculty member late in the agreement and he received UCB-N funding for only the last year.[3]

Figure 8.1 displays the UCB-N funding per year. The UCB-N research committee made a deliberate effort to distribute funding according to merit. As Sydney Kustu, who served on the research committee, put it, "Everyone is concerned about the ways in which such undertakings can go off path. Well, one way to keep them on path is to be sure that all funding is merit based rather than project based." Most of the faculty members considered the procedure by which the research committee disbursed funds to be clear, and people were further mollified because funding appeared to be awarded in an egalitarian manner. Some faculty members thought it was far better to have a committee rather than one person, i.e., the PI, disburse the money, to reduce the potential for favoritism. Many also thought it was a good idea to have the

TABLE 8.1   Number of Benefits of UCB-N Identified by Interviewee Position

Benefit	PMB faculty	CNR faculty, not PMB	UCB administrator	Total
Overhead	14	6	10	30
Funding	20	3	1	24
Seed money	21	0	4	25
Access to information	21	2	2	25
Access to equipment	21	0	1	22
Benefited graduate program	12	0	8	20
Increased productivity	8	0	2	10
Resources	6	0	3	9
Less effort	5	1	0	6
Connection with real world	3	0	4	7
Lack of deliverables	1	0	4	5
Contacts	4	0	0	4
Unified department	4	0	0	4
Staff positions	3	0	1	4

two Novartis representatives on the research committee because this reinforced the sense that UCB-N was a collaborative endeavor and that everyone was working as a team.

BENEFITS OF THE AGREEMENT

During the negotiation phase and early publicizing of UCB-N through newspaper articles and press releases, a number of benefits—to faculty, students, CNR, and the general public—were touted. It is necessary to look at these claimed benefits and determine what occurred in practice, since most of the releases and articles were published before the agreement took effect and thus dealt with hypothetical positives to a large degree.

For Faculty

The PMB faculty believed that their involvement with UCB-N gave them a wide range of benefits. Fourteen different benefits of the agreement were identified by PMB faculty. (see Table 8.1). Two aspects of the agreement appear to have been particularly significant for them: first, money, in one form or another. Fully 38 percent of the benefits PMB faculty singled out were seed money for future projects, general funding, and overhead recovery. Second, NADI made available certain proprietary and confidential databases, permitted faculty to use expensive equipment, and provided expertise unavailable from other sources. These accounted for 34 percent of the claimed benefits, according to the faculty we interviewed.

For Postdoctoral Researchers

Most people said that salary was the greatest beneficial outcome of UCB-N for postdoctoral researchers. With the funding of faculty members' grant proposals came both the need and ability to hire postdocs to carry out the actual research. Indeed, of those faculty members who employed people to work on UCB-N projects, most preferred to hire postdocs over graduate students. At least ten PMB faculty members used UCB-N funds to hire postdocs. While some people were hired to work on a specific project funded by a specific source, others had more autonomy in what they worked on and could be funded simultaneously through multiple grants. This makes it very difficult to ascertain whose job was a direct consequence of UCB-N; in many cases the postdoctoral researchers did not know the source(s) of their funding. As one former postdoc explained, "It's always difficult to say who was directly funded by the agreement, at least in our case. Money came in, several people were hired because there was money available, and now several people have gone because money is *not* available. But we certainly didn't think in terms of 'Oh, you're on the Novartis money and you're not.'"

One longtime member of PMB estimated that there was a twofold increase in the numbers of postdocs in PMB as a result of UCB-N's providing salaries for one or two people for each laboratory. Those postdocs who did work on UCB-N projects also had the benefit of access to the same proprietary databases and technologies used by the faculty. One researcher explained that while the technology resources available through UCB-N were also available to other academics, UCB-N had the huge advantage of having them all together so that work could move more quickly.

For Graduate Students

The desire to improve PMB graduate student education was one of the reasons given by the PMB faculty for entering the agreement with Novartis. In particular, Wilhelm Gruissem, the originator of the search for industrial partnerships, is credited with this desire as his primary motivation. By and large the benefits of UCB-N for graduate students, as identified by the students themselves, mirrored those of the postdoctoral researchers, with money again seen as the primary advantage. PMB graduate students were the beneficiaries of Novartis funds in two ways. First, benefits arrived directly through the $500,000 annually allocated from the Novartis indirect costs to graduate fellowships. The fellowship money was pooled with other fellowship funds so that no particular fellow was directly supported solely with Novartis funds (Price and Goldman 2002).

However, the $500,000 was divided evenly between the plant biology and microbial biology divisions, the latter of which also includes graduate students whose advisors are in other departments such as MCB or Public Health[4] but whose stipend for the first three (later reduced to two) years was nonetheless paid by PMB. The amount of the graduate student stipend increased dramatically, from $15,700 in 1998 to $22,000 in 2003, a rise of 40 percent, an important consequence of UCB-N. ESPM stipends also increased, however, by 37 percent, to $17,427; nutritional sciences and toxicology (NST) stipends increased 40 percent, to $21,000; and MCB stipends increased 44 percent, to $24,500. These figures disprove the claim that PMB graduate stipends rose purely as a function of UCB-N. The second array of benefits was more indirect, arriving through research assistantships funded by way of the students' laboratory affiliations. Intriguingly, this second route appears to have been negligible, as few faculty members reported hiring graduate students with NADI funds or even having them work on NADI-supported research in a capacity other than research assistant.

According to Scott Kroken, a former PMB graduate student, Gruissem had put heavy emphasis on the benefits that UCB-N would bring graduate students. In particular, they were told that a proportion of the funds would be directed specifically for graduate support, "and then we found out what a miniscule proportion that was, and that was one thing that made us quite cynical." In fact, the benefits to graduate students really depended on who their individual advisor was and what faculty members chose to do with their share of the money. For example, one faculty member bought new computers for all the laboratory staff, postdoctoral researchers, and graduate students, who had previously been sharing computers. For those few graduate students who did receive direct benefits from UCB-N, Kroken said, it "turned us around from being completely skeptical to 'No, this is great,' because this is really improving our ability to get our work done much more quickly, much more efficiently." Not all graduate students benefited equally, however, because many faculty advisors chose to spend their UCB-N money elsewhere.

In the survey of PMB graduate students conducted in 2000, graduate students reported specific benefits that included new laboratory equipment, graduate stipends/fellowships, funds to participate in conferences, and access to Novartis data (MacLachlan 2000). While one graduate student said that she appreciated the use of equipment bought with UCB-N funds, she joked that the agreement's biggest effect for her was the amount of time she had devoted to interviews about UCB-N.

The faculty considered UCB-N to have benefited the graduate program through funding and access to otherwise inaccessible technologies and information. One faculty member said that the NADI funds provided flexibility in his laboratory budget so that students were able to try things that would have been too expensive otherwise; specifically, almost everything was now sequenced as a first course of action, which produced some interesting results that would not have been revealed otherwise. However, several faculty members said that UCB-N did not have as great a positive impact as it might have. In Kustu's words, "I think it had potentially enormous value for the department and the university in ways that were not fully tapped. I think in terms of student training and innovation, in terms of student training and development, that, from my perspective at least, I didn't see a lot of investment. ... No one thought as deeply or creatively as they might have about how this money could be used to benefit students." That is, above and beyond the funding for fellowships, Kustu said, "I think there could have been greater benefit derived from the point of view of student training."

CONCERNS REGARDING THE AGREEMENT

Although the overwhelming opinion of PMB staff was that UCB-N was a good thing, they did have a few concerns about the agreement at the outset and with respect to its possible long-term consequences. One concern was the effect of Novartis support on the willingness of other sources, both private and public, to grant research funds. At least one faculty member who participated in UCB-N was convinced that his funding from another private company was pulled because of his involvement with Novartis. Other faculty members, however, talked about how they used UCB-N as leverage to get additional funds and stronger commitments of support from industry.

Another issue was how faculty could assure the confidentiality of information while working on multiple projects and warding off the temptation of a "trade-secret" approach. Some faculty dealt with this problem by completely separating personnel and funding for each project; others paid little heed to confidentiality requirements and discussed everything openly in the lab, although there was reticence to discuss projects with outside people. One senior PMB professor, in describing his lab, said, "everybody knows what everybody else is doing and people benefit from interacting with one another. There is absolutely no distinction or secrecy when it comes to the project being funded by the Department of Energy, or being funded by Novartis, or the USDA."

Four faculty members said that at the start of the agreement they were somewhat concerned that Novartis would be domineering and would try to direct their research to particular questions or delay publication of their results. All four said their concerns had not been realized. Postdoctoral researchers concurred that while some people were convinced that industry involvement would lead to ruin, UCB-N did not work out that way. One PMB faculty member, Renee Sung, quipped that it probably helped that NADI was so far away and thus unable to affect departmental life directly. Unlike people outside PMB, those within the department were not particularly worried about the scope or scale of the agreement. One faculty member said he was initially anxious about the potential for change in the internal structure of PMB, but that this concern too had turned out to be unfounded.

By and large the PMB faculty had no major concerns regarding UCB-N. This may be because, unlike other stakeholders, PMB faculty members had the opportunity to read the draft contract and request the alteration of particular points they disliked. Some graduate students and postdoctoral researchers, by contrast, expressed dismay at their exclusion from the drafting process, as we have seen. But these concerns largely dissipated after the agreement was signed.

CONSEQUENCES OF THE AGREEMENT

Effects on Research Direction

Because of the benefits most valued by faculty (see Table 8.1), many people argued that the focus of their research moved more quickly in new directions than would otherwise have been possible. Almost to a person, the faculty of PMB said that primarily the funds from NADI, and secondarily the access to equipment and information, enabled them to explore research questions that they otherwise would not have pursued or that they would have postponed until they had sufficient results to support a government grant proposal. One junior faculty member said, "It [UCB-N] provided me the opportunity to explore some areas that I otherwise wouldn't have explored, had I not had this money." Several senior faculty members explained how their UCB-N projects in new areas provided the preliminary results necessary for successful grant applications to the NSF, NIH, and other funding agencies.

While it was repeatedly emphasized that these shifts in direction were not dictated by Novartis, the faculty clearly acknowledged that many of the changes would not have occurred without UCB-N. One person was able to apply the new technologies made available by NADI to work he had done fifteen years earlier. This work has generated new results, but he said he never would have gone in this direction without UCB-N. A senior faculty member described the research drift as a subtle transformation:

We did have control. It [UCB-N] didn't change our research; it didn't force us to change our research interests—certainly it changed, because we had the opportunity to look at the information that Syngenta/TMRI have, and that's changed a lot of people's research; there's no question about that. But it wasn't, "You must work on this area," or even, "If you want your proposal funded, it'd better be something that TMRI's interested in." That was not the case at all.

Indeed, many faculty members stressed that the research questions in which they are interested did not change per se but rather that the method of answering those questions changed with access to NADI's technologies. An assistant professor observed, "Research projects always evolve along the way, depending on the availability of the technologies, and it's the same here. Our project wouldn't be the same if there was no such technology available." Faculty thus tended to shift their choice of model organism to one that was sequenced by NADI, e.g., rice. But this may not be much different from a similar shift that occurred in the late 1980s when many plant biologists switched to *Arabidopsis* as a model plant.

Productivity

The increased productivity of PMB was one argument used consistently both as justification for entering UCB-N and also as a measure of the agreement's success. Price and Goldman (2002) examined the participants' curricula vitae and illustrated this increase in publications and grants. There are a number of possible reasons for this increase: (1) UCB-N enabled PMB faculty to continue their current productivity with less effort; (2) UCB-N enabled PMB faculty to do more of what they were already doing; (3) UCB-N enabled PMB faculty to do what they were doing more rapidly, and this enabled them to get to the next project in their research program more quickly; and (4) UCB-N enabled PMB faculty to do some novel, risky things that they would not otherwise have done. In many cases the results from experiments funded by NADI were used successfully as supporting justification in later grant proposals to government agencies such as NIH and NSF. However, the increase in extramural grants within PMB must be seen in the context of a rapid rise in funding for molecular biology nationwide during the period in question (see next section). It is thus unclear how much of the funding increase can be attributed to seed money and other benefits provided by UCB-N.

While those who were central to the creation of the agreement clearly intended the academic participants to strike out in new directions, to try something risky, at least one faculty member was deeply concerned "that the research productivity will not be commensurate with the luxury allowed by funding." Indeed, another faculty member said that he did not explore a completely new direction but rather that NADI funds accelerated the success of his existing work because he could now hire a postdoctoral researcher directly instead of bootlegging the salary from existing projects. Finally, while it is likely that access to Novartis's technologies and proprietary databases had *some* impact on the range of opportunities PMB faculty could explore, we found no meaningful evidence that researchers notably altered their research programs in a fashion that might serve Novartis's interests. Furthermore, had we specifically sought to find such a skewing, the number of intervening variables and the lack of a control group would have made any such determination deeply problematic.

One difficulty with a quantitative productivity analysis is that, if people are truly moving into new areas, then there might well be a lag in publication. As one faculty member said, "My expectation is that Novartis money will lead to *new* (knock-your-socks-off, not same-old) discovery beginning about now [January 2004], but that overall 'productivity' as measured by pubs reflecting the same-old should go down a bit."

Typically, as faculty research programs grew, so did the size of their laboratories, and NADI funds were used to pay the salaries of postdoctoral researchers and technicians. These increases in research programs, laboratory facilities, and staff led to unanticipated restructuring, as the faculty were now responsible for securing larger grants, from either competitive sources or additional contracts, in order to maintain their larger laboratories. Alternatively, they were forced to decide which personnel to lay off or which programs to drop as their laboratories decreased to their former size at the end of the agreement.

Changes in Funding Patterns

There is little doubt that the extramural funds available to PMB increased markedly after UCB-N was signed. The increase is especially striking when one compares the average funds received during the four years preceding receipt of the grant (1995–1998) with the four years following receipt (1999–2002) (Table 8.2). Overall, PMB funding increased by 178 percent during the grant period, and even if one excludes industry funding, funding increased by a hefty 112 percent. Given that the tenured and tenure-track faculty full-time equivalents (FTEs) did not change over that period, this also represents a 178 percent increase in funds per FTE.

TABLE 8.2   Funds Received by Selected Bioscience Departments, 1995–1998, 1999–2002

	Annual average 1995–1998	Annual average 1999–2002	% Change
Plant and Microbial Biology (PMB)			
Federal	$5,382,299	$9,704,123	+80.3
Industry	$816,379	$6,511,325	+697.6
Non–federal governmental	$99,110	$115,133	+16.2
Not for profit	$729,106	$3,173,600	+335.3
University of California	$201,874	$571,354	+183.0
Total without industry	$6,412,390	$13,564,209	+111.5
TOTAL	$7,228,769	$20,075,534	+177.7
Number of FTEs	26.0	26.0	
Funds per FTE	$278,030	$772,136	+177.7
Molecular and Cell Biology (MCB)			
Federal	$30,013,700	$44,965,432	+49.8
Industry	$290,526	$473,686	+63.0
Non–federal governmental	$100,91	$270,278	+167.8
Not for profit	$2,831,019	$6,383,044	+125.5
University of California	$635,706	$639,044	+0.5
Total without industry	$33,581,336	$52,257,798	+55.6
TOTAL	$33,871,862	$52,731,484	+55.7
Number of FTEs	70.5	70.0	
Funds per FTE	$480,452	$753,307	+56.8
Plant Biology at MSU			
Federal	$1,365,795	$2,262,492	+65.7
Industry	$448,514	$2,519,629	+461.8
Non–federal governmental	$65,952	$15,131	−77.1
Not for profit	$344,954	$150,287	−56.4
University of California	$245,573	$328,618	+33.8
Total without industry	$2,018,673	$2,754,777	+36.5
TOTAL	$2,467,186	$5,274,407	+213.8
Number of FTEs	30.5	24.6	
Funds per FTE	$80,891	$214,407	+265.1

Sources: UCB SPO; MSU, Contracts and Grants Administration.

Given the similarity in subject matter between PMB and MCB, it is also instructive to compare the two departments. As noted in Table 8.2, MCB increased its total funding during the grant period by 55.7 percent. When industry dollars are not included, the numbers differ only slightly, as industry funds made up only a small portion of MCB's extramural research portfolio (<1 percent) as compared to 32 percent of PMB's portfolio. Here, too, faculty FTEs changed only slightly during this period, so the increase in funds per FTE is also 55.7 percent.

Furthermore, in terms of dollars per FTE, the considerable disparity between the two departments in the earlier period has disappeared, such that during the grant period MCB funding per FTE was actually slightly lower than that of PMB. In the pre-grant period, PMB faculty FTEs had only 58 percent of the funds that MCB faculty had; during the grant period they had 102 percent as much research support as their MCB colleagues. In short, for all practical purposes we can say that the extramural support per FTE received by the end of the UCB-N agreement was about the same in both departments.

To put these figures into perspective, it is instructive to compare PMB with another plant biology unit. Given ease of access to data, we compared it to Michigan State University's Department of Plant Biology.[5] That department, like PMB, does relatively little applied work and receives relatively little state or university funding for research. There we found that, in part as a result of considerable industry investment (a grant of $9.5 million to several faculty members in FY 1999), funding in the 1999–2002 period was up 214 percent. This works out to a growth per FTE of 265 percent. Moreover, even when industry funds are not included, funding rose 37 percent across the two periods—and this in spite of a considerable loss of FTEs. Also worthy of note is that federal grant funding rose significantly across the several time periods in all three cases: 80 percent for PMB, 50 percent for MCB, and 66 percent for MSU's Plant Biology Department. In sum, the argument that the Novartis agreement was a necessary condition for enhanced funding of research in plant biology is tenuous, given the increased amount of federal and other funds available for work in this area.

Furthermore, while PMB faculty and CNR administrators complained of declining public funding opportunities, the data do not quite support that interpretation of the situation. True, the federal *formula* funding received by the Division of Agriculture and Natural Resources (DANR) has been stagnant over the past decade. Similarly, federal extension funding (also based on a formula) has declined. Of course, as was true in many states, the state of California did have its own fiscal crisis in the early 1990s, leading to declining state support of higher education.[6] The university raised tuition and cut programs as a result of the deficit. But California state funds for research were and remain an insignificant part of PMB's extramural funds. In contrast, other federal funds for academic R&D have actually continued to rise over the entire postwar period (with a few short-term declines). When one looks more carefully at funds for life sciences research, one finds a similar situation.

As Figure 8.2 shows, although biological research has not received the government largesse given the medical sciences, federally financed biological

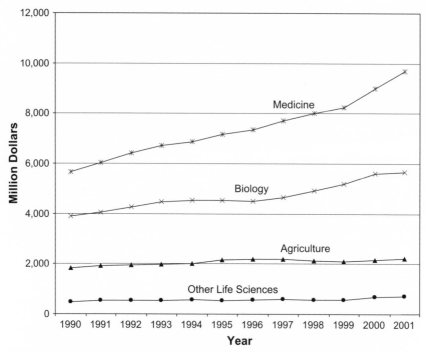

FIGURE 8.2   Expenditures for life science academic R&D, 1990–2001 (constant dollars). *Source:* National Science Board (2004), "Deflators from National Research Council, 2002."

research has gone up dramatically over the past decade, from about $4 billion to nearly $6 billion. This represents a 50 percent increase in funding in constant dollars. It might be argued that such figures are inappropriate, as the research performed in PMB is agricultural in nature. Indeed, federal support for agricultural research has increased much more slowly. But PMB faculty frequently noted in interviews that their research was not directly connected to agriculture. They stressed that they engaged in basic research in biology, far removed from the applied work of their colleagues at UC Davis and UC Riverside. If that is the case, then the national biological research funding trend is the most appropriate for consideration when examining PMB. It in no way suggests a funding shortage.

Instead, every time the federal government increases the number of research dollars, the number of universities and researchers competitively pursuing those dollars increases even faster. Indeed, the proportion of total research dollars going to the nation's top universities has declined as the absolute size of the budget has increased. This drives the best universities, despite their greater competitiveness, toward corporate collaboration. More and more faculty, staff, and administrators spend more and more time preparing, processing, and reviewing unsuccessful grant applications.

In economic terms, the transaction costs rise rather than fall as the pie gets larger. Since industry funding is usually based on different criteria then federal research grants, top-flight researchers and administrators may look in that direction in their search for new funds. Furthermore, the incentive and reward systems at institutions historically focused on teaching shift toward grantsmanship and away from their historical commitments (Washburn 2005).

The British government, faced with the same dilemma, recently decided to refuse to fund all but elite university research programs. This approach, of course, means that excellent researchers at weaker institutions are denied access to research support. This may or may not have negative consequences for the progress of science and society at large. Regardless, the current situation puts research universities on a never-ending treadmill, as they must work ever harder to secure a declining share of federal research dollars.

Effects on Graduate Students

Many UCB faculty were concerned about the potential effects of UCB-N on graduate student education in PMB. But the direction of graduate student research appears not to have been altered by UCB-N, and 60 percent of students surveyed had a positive assessment of the agreement itself and the attendant resources (MacLachlan 2000). The views held by graduate students on the negotiation and implementation of UCB-N have already been discussed in the relevant subsections above. In this section we look at the effects of UCB-N on the size of the graduate student cohort, support for graduate students, and internal relations.

The most significant consequence of the agreement for the students of PMB is that it enabled the graduate program to increase in size and annual stipend amount. PMB was able to double the size of its graduate program; the number of applications for admission, acceptances by the department, and enrollees all increased. It is less clear whether PMB has become more attractive as a graduate department relative to other molecular biology departments. On the one hand, the number of applications increased by more than 100 percent in three or four years (see Figure 8.3), which suggests that some potential students were applying to PMB who would not previously have done so. However, the percentage of applicants accepted by the department who then chose to enroll in PMB did not increase. This challenges the argument that UCB-N enabled PMB to attract graduate students who otherwise might have chosen (rather than been forced to enter) a different program.

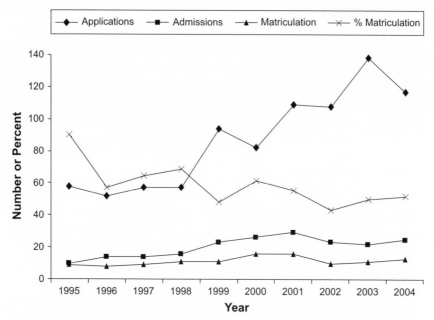

FIGURE 8.3   PMB graduate student applications, admissions, matriculation, 1995–2004. *Source:* PMB enrollment statistics.

Just as the larger laboratories made possible by UCB-N funds had unanticipated drawbacks, so too did increasing the size of the graduate program. For the cohorts entering during the first four years of the agreement, PMB, as a department, promised to support the first three years of each graduate student's stipend. At the end of the Novartis agreement, however, departmental support decreased to the first two years, leaving the faculty and/or student to fund the remainder.

There is a marked difference in attitude about UCB-N between graduate students who enrolled in the program before or during the UCB-N negotiations and those who joined PMB after 1998 (MacLachlan 2000; Price and Goldman 2002). Students who entered after 1998 were far less concerned about both the lack of graduate student involvement in negotiations and the terms of the agreement itself. This split may indicate a normalization of university-industry relations, in that the benefits and consequences of such relations are now less subject to question. In the words of one PMB graduate student, "It is a pretty reasonable agreement given that the university would end up selling commercial inventions and discoveries anyway. Everyone signs patent agreements anyway, details don't matter so much" (quoted in MacLachlan 2000, 5).

Or the split may reflect the fact that students entering after 1998 applied to the program at least in part *because* of the agreement, while students already there when the negotiations began did not. The internal review of UCB-N conducted by the office of the vice chancellor for research also took note of this split: "New graduate students appear to have only the faintest idea about the Novartis Agreement, indicating that it is not a major matter of discussion within PMB" (Price and Goldman 2002, 38). Since many of today's graduate students are tomorrow's professors, this shift in expectations is especially important. How graduate students are professionalized has a direct bearing on their actions as professors.

Undergraduate Education

It is very difficult to compare the teaching responsibilities of individuals across campus before, during, and after the five-year period of the contract, though it is possible to examine the number of students who have declared PMB as their major and the number of regularly scheduled classes offered by PMB over time.

UCB-N was signed almost halfway through the 1998–99 academic year. If this year is removed from the calculations, then in the five academic years prior to UCB-N, PMB offered an average of 23.75 classes per year. The five academic years during which UCB-N was in effect show an average of 27.5 classes per year. Of course, these classes were not all taught by tenured and tenure-track faculty; they include classes taught by adjunct faculty and temporary lecturers. When the same calculations are done for the number of declared undergraduate majors in PMB, we see that prior to the agreement an average of forty-five undergraduates were PMB majors each year. Immediately after UCB-N was signed, and possibly reflecting the amount of negative media attention, the number of majors declined, although it then increased dramatically three years later, such that on average the number of PMB majors in the last two years of the contract was again forty-five. There was no similar decline in the number of majors in the other departments in CNR or in MCB. This suggests that the negative publicity surrounding UCB-N did have the temporary effect of depressing the number of undergraduates who chose to major in PMB.

One of the concerns of the Academic Senate was that PMB faculty members would use the Novartis grant funds to buy out their teaching commitments. This does not appear to have happened. Zacheus Cande, an MCB faculty member with a joint appointment in PMB, explained that the disparate teaching loads are a point of tension between PMB and MCB. MCB faculty members "have more teaching responsibilities than we can necessarily dredge up faculty members for; whereas plant biology [PMB] has just the opposite problem—they have more FTEs than the amount of teaching would justify." In Cande's opinion, PMB has a small number of undergraduate majors because the pre-med students have a prejudice against plants, see them as irrelevant to their medical career, and thus choose to major in MCB rather than PMB. Since the teaching budget is distributed according to how many students are enrolled in departmental classes, only a small portion of the PMB budget is supported by university general funds.

When asked about the benefits of UCB-N for PMB undergraduates, PMB faculty members were hard pressed to come up with any. Most said that they did not use the information from the Novartis databases or the equipment bought with Novartis funds in teaching undergraduates, although several said they referred to Novartis's equipment and information as illustrations of what could be done. The most direct benefit for undergraduate students is in Sydney Kustu's laboratory, where she teaches undergraduates how to do *E. coli* arrays. Kustu is able to do this because UCB-N funded her to do genomic arrays in her own laboratory rather than send the samples elsewhere for analysis. Although the benefits are minor, no one was able to identify any negative effects on undergraduate education.

Effect on Internal and External Relations

One indicator of healthy diversity is collegiality. PMB faculty members reported that collegiality had been enhanced at least somewhat within the department. For example, annual retreats with NADI brought nearly the entire PMB faculty together, whereas previously there was little interaction between the plant and microbial divisions within the department. But if these annual retreats increased interaction, then those who were not part of UCB-N lost out on opportunities for collegial association. There was no consensus regarding the effect of UCB-N on collegiality within CNR because some people see the agreement as a divisive lightning rod that exacerbated preexisting conflicts, while others see no residual tension. Several faculty members mentioned that colleagues at other universities asked how the agreement was working out when they met at professional conferences. But most PMB faculty thought these questions were motivated by envy rather than disapproval.

Along similar lines, the reputation of PMB appears not to have been negatively affected by the agreement with Novartis, although people seem to distinguish between issues of reputation and issues of bias. That is, while observers have tended not to see the reputation of PMB and its faculty sullied by UCB-N, they do tend to approach PMB research with skepticism. One PMB faculty member said he cannot control how other people view PMB's ties with industry and that this "loss of perceived purity" is the worst consequence of industrial funding. Some graduate students also believed that UCB-N had negative consequences for the perceived objectivity of their research. One PMB graduate student said, "[I] don't think it's such a bad deal, except in terms of the public perception, which might be detrimental to the department as well as public opinion about science. The public thinks that the department has sold out" (quoted in MacLachlan 2000, 4). We shall return to this point for UCB as a whole in Chapter 10. Our information about how outsiders view the objectivity or PMB research performed under UCB-N is entirely anecdotal, however, and cannot be measured empirically.

NINE

Intellectual Property Rights

ALTHOUGH INTELLECTUAL PROPERTY rights (IPR) are mentioned specifically in the U.S. Constitution, there was widespread agreement in the late nineteenth century that they were inapplicable to plants and animals. Bacteria and other microorganisms were considered patentable, but there was a general agreement that higher plants and their component parts should not be subject to utility patents. This consensus was based on the four requirements for a utility patent: novelty, non-obviousness, utility, and specification.[1]

In legal terms, "novelty" is usually taken to mean a novel composition of matter. Thus discoveries are normally not patentable as they exist already and are merely made known by the discoverer. In practice, the demonstration of novelty is usually met by noting that the object in question has not been patented before, is not in general use already, and is not easily produced by someone already skilled in that particular art. "Non-obviousness" means that the thing to be patented requires some inventive step beyond what skilled practitioners usually do. "Utility" is broadly interpreted as the demonstration of some public benefit to be had from the invention. Finally, "specification" involves a written description of the item that could be employed by anyone knowledgeable in the field to create the object to be patented.

Plant materials were excluded from utility patents on several grounds, including (1) the impossibility of providing an adequate specification, (2) the lack of invention (i.e., the claim that novel plant materials are essentially a discovery), and (3) the lack of novelty (i.e., anyone skilled in the art could produce the item in question). As a result, private investment in plant improvement was relatively unprofitable throughout the nineteenth century and much of the twentieth. After all, since seeds are self-replicating, they can be copied and freely grown by farmers, other seed companies, or public-sector research institutions such as LGUs. Indeed, all three of these groups copied and grew seeds well into the twentieth century; therefore, seed prices remained only slightly above grain prices (Kloppenburg 1988).

Hybrids, developed in the 1930s, presented a partial solution to the patent problem for seed companies by creating the biological equivalent of a patent. Since the progeny of a hybrid plant had little or no yield when planted, farmers had to return to the marketplace each year for more seed. And because seeds were a relatively small portion of the total cost of farming, they were willing to pay the higher price for the benefits of hybrids, including improved yield. The hybrid seed business soon boomed, although for technical reasons it was limited largely to hybrid corn.

The Plant Patent Act (PPA) of 1930 (*U.S. Code* 35 [1930], §§ 161–64), championed by Luther Burbank and of particular interest to flower and fruit growers, was the first attempt to provide patent protection for plant matter. PPA permitted the patenting of clonally reproduced plants (i.e., reproduced from cuttings and tubers). Unlike the more stringent requirements for utility patents, filers of plant patents did not have to show how to produce the specific variety. They only needed to file a description that was "as complete as possible." Moreover, since creating a plant *de novo* from other matter was and remains impossible, people and companies seeking plant patents were required to submit a sample to a public repository. Finally, the scope of the claim was limited to the variety of plant for which the patent was sought. In addition, plants produced outside the United States using the same plant matter could be imported without infringing the rights of the patent holder. In sum, PPA provided certain protections to "inventors," but it was more limited in scope than the laws protecting utility patents.

For the next forty years there were few changes in IPR concerning plants. Then, in response to considerable pressure from the seed industry, Congress passed the Plant Variety Protection Act (PVPA) (*U.S. Code* 7 [1970] §§ 2321–2582). The argument for the act was based on the highly debatable notion of "plant breeders' rights." In point of fact, by 1970 there were few independent plant breeders in the United States; most were employees either of State Agricultural Experiment Stations or of private seed companies. At the time, most public breeders were opposed to the new law, seeing it as the privatization of a public good (Fejer 1966).

The new law extended patent-like protection to most sexually propagated plants provided that they met the three criteria of novelty, uniformity, and stability. In other words, they had to have novel characteristics as compared to other plants of the same species, they had to express that novelty uniformly over space, and they had to be stable through time (i.e., they had to breed true). Unlike utility patents, PVPA did not limit novelty to useful differences; until later amendments in 1991, minor cosmetic differences were sufficient to permit the granting of a certificate. In addition, unlike utility patents, PVP certificates could be issued even if someone skilled in the art was also able to produce the new variety. Finally, the PVP certificate was issued for the concrete object, not for the idea behind it. Those granted a certificate could bar others from selling the variety for a period of twenty years.

For the next ten years little else happened. Then, in 1980, a Supreme Court decision in the case of *Diamond v. Chakrabarty* (447 U.S. 303) opened the door to utility patents for plants. The government argued that the very existence of PPA and PVPA demonstrated clearly the inapplicability of utility patents to life forms: Life forms were not created by people but were found in nature. UC filed a brief in support of extending patents to life forms, arguing,

> Whether the University has the right to patent its own newly manufactured microorganisms will depend directly on the disposition that is made in this case. In turn, this will govern whether the University receives income from these inventions, to be significantly shared with its inventors and to use, *inter alia,* in supporting new research. Indeed, if no patents issue, the health care industry may well elect not to commercialize these important inventions because of its avowed belief that absent the protection a patent affords, the time and experimental work requisite to obtaining government clearances cannot be justified. (Brief dated January 28, 1980, *Diamond v. Chakrabarty,* 447 U.S. 303 [1980])

In essence, the court ruled that life forms could be patented if they were made by humans. Although the case involved a microorganism, it was soon applied to all life forms. Soon thereafter, the U.S. Patent Office began to issue patents for plants. Within biotechnology, an expanded range of IP was now protected, including genes, promoters, transformation processes, plant germplasm, and new plant varieties. In 2001 the Supreme Court clarified the distinction between utility patents and the PVPA for agricultural seeds. The Court explained:

Because it is harder to qualify for a utility patent than for a Plant Variety Protection (PVP) certificate, it only makes sense that utility patents would confer a greater scope of protection. … The PVPA … contains exemptions for saving seed and for research. A farmer who legally purchases and plants a protected variety can save the seed from these plants for replanting on his own farm. … In addition, a protected variety may be used for research. … The utility patent statute does not contain similar exemptions. (*J. E. M. Ag Supply, Inc. v. Pioneer Hi-Bred International, Inc.,* 200 F.3d 1374, 213 and 215 [Fed. Cir. 2000])

In sum, in approximately half a century, U.S. law went from a complete prohibition on utility patents for plants to measures that encouraged the creation of such patents. This had a profound effect on the seed industry as well as on universities. Private-sector breeding of major crops increased dramatically, while public-sector breeding declined. As breeders retired they were replaced by plant biologists with training in the use and development of the new biotechnologies (Frey 1996; Knight 2003).

With the cultivation of IPR, companies captured commercial value from the new science. Firms may use several different techniques to capture both codified and tacit knowledge necessary to exploit scientific discovery (Quintella 1993). Companies may apply for a patent based on the results of internal R&D, although a patent is no guarantee that an exclusive right can be maintained. A firm must have the financial reserves to sustain an active R&D program, the ability to continue filing and prosecuting patents in a fast-moving field, and significant financial resources to defend its patent rights through litigation if necessary. In addition, companies may need to acquire IP by licensing technology from third parties. A company may obtain IP by licensing it from another company or university. When two or more parties have IP of interest, cross licensing of patents may be needed for each participant to have freedom to practice the new technology. DuPont and Monsanto, for example, reached cross-licensing agreements for biotechnology products in corn, canola, and soybeans in 2002 (Thayer 2002). Merging with or acquiring a firm already owning IP can be an additional strategy. To be effective, strategic decisions must also reach beyond patent protection to consider the availability of know-how, trade secrets, market access, and tangible assets.

Inserting the management of IPR into academic administration has been a work in progress for almost a century. The formal story of university technology transfer begins with UCB chemistry professor Frederick Gardner Cottrell's interest in commercializing technology for the electrostatic precipitation of fine particulate materials. Unable to persuade university administrators to assist him directly, Cottrell labored to develop an independent organization. This effort led to the establishment of the Research Corporation in 1912. Similarly, the University of Wisconsin chose to develop an independent organization to manage IPR. The Wisconsin Alumni Research Foundation (WARF) was formed in 1925 (Marcy 1978).

Despite noteworthy successes by the Research Corporation and WARF, universities moved slowly to capture IPR. UC did make the assignment of faculty inventions mandatory in 1963 (Owens 1978), but the process of technology transfer at universities remained slow and halting until the Patent and Trademark Laws Amendments of 1980, which included the Bayh-Dole Act (Bremer 2001). This act streamlined the process by which universities, as well as other nonprofits, small businesses, and contractors operating government laboratories may retain title to inventions funded by the federal government. The importance of university IPR in the field of biotechnology became clear in 1981 with the nonexclusive licensing by Stanford University of the Cohen and Boyer patents (U.S. Patent nos. 4,237,224, 4,468,464, and 4,740,470, all filed on November 4, 1974). Even with this activity, most universities did not have separate offices for handling technology transfer until the 1990s. Equity transactions have an even shorter history. UC, for example, released its "University Policy on Accepting Equity When Licensing University Technology" in February 1996. University policy and practice regarding technology transfer of university-owned IPR continue to evolve through an active process of learning by doing.

Within the university setting, technology transfer offices are responsible for processing invention disclosures received from university researchers. The available pool is assessed for patentability and for its value if successfully licensed. The tangible output of a technology transfer office includes a number of standard agreements crafted to define and manage IPR, such as (1) confidential disclosure agreements, also known as non-disclosure agreements or secrecy agreements, (2) material transfer agreements (defining rights to tangible property transferred between two parties), (3) research agreements, (4) option agreements, (5) license agreements, and (6) inter-institutional agreements.

IP agreements are critical points of contact between universities and industry. In preparing these agreements, the technology transfer office tries to strike a balance between the multiple missions of an LGU relative to IP. At UCB the decisions are guided by established policies, including (1) patent policy, (2) copyright policy, (3) policy on accepting equity when licensing university technology, (4) guidelines on UIR, (5) guidelines on licensing, and (6) conflict of commitment and outside activities of faculty members. Broader policy, such as codes of conduct, may also serve to regulate technology transfer activity.

With successful commercialization of technology protected by IPR, monopoly rents may be legally collected.[2] The sale of products and services returns money through rents only after legal and regulatory hurdles are cleared. To profit from research generated and IPR, licensees of the new technology must work out an internal or external manufacturing and distribution system to reach the end user. The licensee can set a higher price for its goods based on market conditions and the strength of its IPR and related assets. With this accomplished, rents filter up through the distribution channels to the private-sector licensee. Assuming that the new technology may be produced at or below the cost of competitive products, monopoly rents may be used to deliver profits for shareholders and support R&D for future product introductions. Licensees of these patents provide scale-up of the discovery, combine them with other IPR, conduct field trials, seek regulatory approval, and distribute and market the product to end users.

The licensee of university-owned technology returns a royalty based on revenue generated from its rents. The university will reinvest a portion of these funds into additional research of its choosing. Faculty members may also benefit directly; individual researchers may keep a share of the proceeds but may also divert some or all revenues back to research. Proceeds may also flow to a startup company. Money from these firms may then be transferred, to the benefit of venture capitalists, with a portion due to the university for technologies licensed. Again, the faculty member may benefit directly from the startup's share. In other instances, the startup company may already have been acquired by the time a commercial product is generating income. In this case the venture capitalists may have already cashed out, returning money to their investors. Finally, compensation may be due to the faculty member from the university, depending on its patent policy.

The Association of University Technology Managers, Inc., has conducted national surveys since 1991 on patenting, licensing, research, and other aspects of technology transfer for American and Canadian universities, research centers, and hospitals (AUTM 1995–2003). The number of invention disclosures, patent applications filed, and patents issued all increased significantly throughout the 1990s. In line with national trends, technology transfer at UC expanded consistently throughout that decade. UC experienced increases in the number of inventions reported, patents filed, and patents issued. The number of inventions reported by UC rose from 500 in 1992 to 957 in 2001, an increase of 91 percent.

UC is one of the leaders in technology transfer. In 2001 it produced more inventions, had more patents granted to it, and commercialized inventions more successfully than any other university. No institution has a longer or more successful track record in technology transfer than UC. A recent UC (2002, 13) report finds that the "University technology transfer program continues to expand and flourish. ... In 2001, 957 inventions were disclosed, a 45% increase over the 661 inventions disclosed in 1996. Over the same five-year period, the total number of active U.S. patents increased by 100 percent and the overall portfolio of patents by 75 percent. Total licenses and options held by The Regents increased by 65 percent. Since 1996 the University has earned almost $632 million in income from royalties and fees."

Moreover, *Technology Review*'s university research scorecard for 2001 ranked UC number one for the strength of its patent portfolio. The other nine universities in the top ten for 2001 were MIT, Stanford, Caltech, University of Texas, University of Washington, University of Wisconsin, Columbia University, University of Michigan, and Johns Hopkins University (Brody 2001). The Berkeley campus saw significant increases in its technology transfer program during the same period. The number of inventions, patent applications, and patents issued per year all increased significantly. For FY 2001, UCB OTL and its faculty inventors netted $4.3 million after payments to joint IPR owners and legal expenses. While substantial, this may be compared to total UCB expenditures in the same year of $1.3 billion and a campus endowment of nearly $2 billion.

Perhaps the most ironic twist with respect to the growth in IP on university campuses is the recent series of court decisions in *John M. J. Madey v. Duke University*. The relevant part of this case is that Duke University claimed that it had used a particular patented instrument solely for academic noncommercial research and that therefore it did not need to pay royalties to the inventor. The Federal Circuit Court of Appeals ruled against Duke and for the plaintiff. In making its decision the court noted that

> major research universities ... often sanction and fund research projects with arguably no commercial application whatsoever. However, these projects unmistakably further the institution's legitimate business objectives, including educating and enlightening students and faculty participating in these projects. These projects also serve, for example, to increase the status of the institution and lure lucrative research grants, students and faculty.
>
> In short, regardless of whether a particular institution or entity is engaged in an endeavor for commercial gain, so long as the act is in furtherance of the alleged infringer's legitimate business and is not solely for amusement, to satisfy idle curiosity, or for strictly philosophical inquiry, the act does not qualify for the very narrow and strictly limited experimental

use defense. Moreover, the profit or non-profit status of the user is not determinative. (*Madey v. Duke University,* 307 F.3d 1351, 1362 [Fed. Cir. 2002])

When Duke appealed to the U.S. Supreme Court, UC was a party to an *amicus curiae* brief filed by the Association of American Medical Colleges. The brief explained, "Universities will be forced to bear substantial administrative and financial costs to cover patent searches, infringement options, licensing agreements, and the inevitable litigation that will be engendered by the Federal Court's new rule of patent law" (Keyes and Jones 2001).

Despite the briefs, the Supreme Court rejected the appeal. While Duke University is appealing the case on other grounds, it is likely that in the future universities will incur new costs for the use of patented products in research—costs that will undoubtedly offset many of the financial benefits provided by university-owned patents.

Expectations with respect to PMB's generation of IP have, to date, remained unfulfilled. Few or no benefits, in terms of patent rights or income, to either UCB or Novartis emerged from research conducted during the term of the agreement.[3] Of the fifty-one disclosures made by PMB faculty during the period November 23, 1998, to November 23, 2003, twenty were patented. Ten of these patents were on disclosures funded through UCB-N, at least partially, and NADI optioned three of them. No options to negotiate an exclusive license survived the agreement, however.

Regarding potential licensing activity, TMRI apparently took the greatest interest in work on allergens. One patented technology, for example, suggests possible treatments for allergies, including those triggered by ragweed pollen, with reduced side effects (e.g., anaphylactic reactions).[4] While challenging science, this is also an area of immediate commercial application. In the March 14, 1996, issue of the *New England Journal of Medicine,* researchers at the University of Nebraska, University of Wisconsin, and Pioneer Hi-Bred International, Inc., found that an attempt to transport the higher protein value of the Brazil nut to the soybean resulted in the transfer of the nut's allergenic characteristics. The Pioneer-funded research concluded that "Our study shows that an allergen from a food known to be allergenic can be transferred into another food by genetic engineering" (Nordlee et al. 1996, 688).

Bob Buchanan, as a researcher under the UCB-N agreements, wrote in the journal *Plant Physiology* of the ensuing growth in the research agenda related to allergies: "The development and commercialization of a variety of food crops with transgenes has thrust the allergy issue onto a public stage and given the field unprecedented exposure worldwide. ... The increased public

awareness of food allergy has arisen from a combination of three factors: reasoned concern, fear through ignorance, and political motivation" (2001, 5).
Buchanan's essay spans the boundaries between scientific research, corporate
interests, and regulatory policy:

> One precautionary note seems in order. While proceeding with allergy
> testing, we must be careful not to overregulate and impose undue restric-
> tions to stifle innovation. Rather, we should seek to formulate a balanced
> policy that insures food safety without hindering product development.
> ... I am confident that, with progress now being made, one or more reli-
> able animal models will soon be available to serve as an indicator of aller-
> gens in human[s] and that a safe but reasonable testing policy will be for-
> mulated. Once such testing capability is in hand, the public will respond
> in a positive manner. In the long term, the food allergy and technology
> fields will likely benefit, rather than suffer, from this pause in their devel-
> opment. (2001, 7)

Buchanan was involved in this line of research well before the UCB-N
agreement, with patent filings dating back to 1991, so it appears that the agreement reinforced rather than altered his research agenda in this instance. Over
the years, funding sources for this line of research at Berkeley included NSF,
the Public Health Service, USDA, the BioSTAR Program, Applied Phytologies,
Inc., Coors Brewing Co., Novartis, Inc., Syngenta, Inc., and a gift from Cargill,
Inc. (Hartman et al. 1993; Cho et al. 1999; Besse et al. 1996; Buchanan et al.
1997; Yano et al. 2001).

One widespread concern about UCB-N that came up repeatedly in our
interviews was the scope of the IP to which Novartis had access. From PMB
to members of agricultural commodity groups, a significant number of people expressed uneasiness about giving Novartis the ability to review all of the
research—whether funded by Novartis or by a government or other public
source—conducted by anyone in PMB, as well as the option to negotiate a
license for up to approximately one-third of those discoveries annually.

Many people argued that this provision was at the heart of UCB-N's
novelty and that it potentially had the greatest ramifications, because it
expanded the scope of research beyond the usual arrangement, which allows
companies to negotiate licenses only for research products for which they
paid directly. Indeed, in comparison to gifts and individual faculty contracts—the two predominant ways in which industry disburses money to
public academics—UCB-N is extraordinary in the scope of the rights it gave
Novartis.

Even the former interim dean of CNR, Richard Malkin, who is also a PMB
faculty member, expressed astonishment over UCB-N's scope in this respect:

> There are aspects of the agreement that I think are bizarre; it's the only
> word I can use. That Novartis can come into a laboratory and they have
> access to things, results, and anything that isn't supported by Novartis
> money. I don't understand how the university could sign off on that. That
> is, I have an NIH grant, and I'm working on a project, and then I have
> Novartis money, and Novartis can come in and say, "Well, you know, that
> project over there is pretty interesting. We claim intellectual rights over
> that." When I first heard that at a meeting, I thought, "Well, this just isn't
> right and it can't be something that everybody agrees to," but it was.

Another point related to the IP scope of UCB-N was mentioned by only
a few people, but it is nevertheless important. Novartis's access was limited
to a proportion of the discoveries commensurate with the company's proportion of departmental funding, but Novartis did not have to specify what IP
they were going to pursue until the end of each year, rather than as each discovery was disclosed. In addition, the first three years of the contract were
collapsed into one period from which Novartis could pick and choose the IP
to which it might lay claim. As a senior scientist at PGEC pointed out, "If you
think about it, if you have ten patents, all ten patents cannot be equal. Even
if somebody says, 'I'm only going to take ten percent,' they're going to take
the best out of the ten patents; in other words, that best out of the ten patents
is probably equal [in value] to 99 percent. So essentially it's an agreement that
pretty much says they're going to take the best." This perception squares with
a recent study of patenting at sixty-two research universities, which notes
that the top five inventions accounted for 76 percent of licensing revenues
(Thursby et al. 2001).

Once again, how things worked out in practice did not bear out the concerns expressed before the agreement was implemented. PMB faculty generally reported that they saw nothing especially advantageous to Novartis from
the "peeking rights" to their disclosures. As one associate professor described
the process, "as a first step, OTL asks Novartis if they're interested. If they're
not, then they ask everybody, instead of just asking everybody right off the bat."
Nor did PMB faculty conceive or pursue their research differently as a result
of Novartis's access to IPR, as some had feared. A junior faculty member told
us, "patenting is not the thing that's on the top of my mind with any of the
research we do. If there is a potential application, I would like to see someone
develop it and see whether it really is going to be a useful application of some

basic research. But I certainly don't think about it on a daily basis nor gear my research towards patenting things or anything like that."

Paul Ludden, dean of CNR and also a PMB faculty member, argued more broadly that even though administrators may push for faculty members to disclose and patent inventions whenever possible, there is no provision in the tenure process that explicitly gives extra weight to how much IP a person has patented; therefore this trend is unimportant because it has not been incorporated into faculty's duties. Even so, it must be remarked, patents are important enough as intellectual credentials to be listed on a person's curriculum vita, so any change in this area is worthy of deliberation.

Two senior employees of Syngenta said that, while securing new IP was important because the company is always looking to increase its patent portfolio, it was not the company's primary motivation for entering UCB-N. Indeed, they told us, there were often discrepancies between what PMB faculty considered worth patenting and what Syngenta considered worth the thousands of dollars it costs to patent a new product or process. Another problem with patenting at such an early stage is that discoveries have to be policed in order to defend the patent, which is often not worthwhile when commercial applications may be ten to fifteen years away. In addition, over the five years of UCB-N, the U.S. Patent Office has changed the rules and is now much stricter in requiring applicants to demonstrate commercial utility as part of the patent process (Busch 2002). Given these changes, Simon Bright, head of technology interaction for Syngenta, said that although very few discoveries are taken through the patenting process, "I think that if we were setting out to say, 'Okay, we want to patent in a certain area,' then you just wouldn't do it like this. In that sense what you're trying to do is give smart people the money and see what comes out, and sometimes what comes out is inspiration rather than IP."

As the five-year agreement concluded and all the claims that either Novartis or Syngenta could make were filed, there was no significant IP in evidence. Nevertheless, one "winner" could change this situation and generate significant income for UCB as it moves into commercial application.

As noted earlier, the Bayh-Dole Act of 1980 streamlined numerous government procedures on IP. It provided eligible institutions considerable discretion and does not mandate participation by universities. For example, universities are free to elect or not elect title to each invention arising from research done with federal funds. When licensing IP, the act does call for a small-business preference, but universities are free to license to large firms when they are co-sponsors of research. Universities must also have written arrangements that include financial incentives for campus inventors. Universities control the specific terms of these incentives, as long as income received by the institution is reinvested in education or research.

The PMB faculty we interviewed did not express a great deal of interest in future royalties or equity income that might result from UCB-N. At the same time, however, the need for new research funding was a consistent theme among both scientists and administrators. The financial incentive to work with industry was driven less by the possibility of future returns from IPR than by a perceived lack of government support. Bayh-Dole does not include industry-sponsored research directly, but it does enhance the prospects for private, for-profit firms sponsoring research on campus. As with UCB-N, companies can leverage their own funding by licensing the intellectual output of both their own and government funds.

One of many business-promotion strategies of the 1980s and 1990s, the Bayh-Dole Act had been law for sixteen years before UCB-N. It does not appear to have been a trigger or central cause of the controversy surrounding the agreement. Indeed, Richard Nelson (2001) has argued that Bayh-Dole merely reinforced trends that began long before the act was passed. Consider, for example, the impact of the changes in patenting of plants and plant parts brought about by both positive law and case law noted above. Moreover, as Nelson rightly notes, Bayh-Dole may serve to gradually shift faculty interest from publication to patents, thereby undermining a central value of academic science, and may give the public the erroneous impression that universities are in business to make money rather than promote science in the public interest. Moreover, the shift in the role of IP and the controversy engendered in part by this shift do demonstrate the need to formulate university policies more carefully and to better communicate the nature and approach of IP management within UCB and among the general public.

SURROUNDING CONTROVERSY

Most PMB faculty members thought that the controversy surrounding UCB-N stemmed from the preexisting division in CNR between those who work on molecular phenomena and those who work on ecological phenomena. Rausser's extension of the agreement to CNR exacerbated this division, as did the general perceptions of the lack of transparency. If the agreement had been restricted in name as well as practice to PMB, then, faculty members believed, there would have been less criticism. As one senior faculty member said bluntly, "I would say, 'It's none of your damned business; go get lost; I do whatever I like to support my research,' ... and it would have ended there."

Instead, people outside PMB were associated with UCB-N without being involved in the decision-making process.

Most people in PMB understood why other CNR members resented the whole college being dragged into something they thought was wrong, especially when CNR as a whole was prevented from participating in the negotiations and did not get the same benefits. If the agreement had been with PMB alone, it would not have threatened the reputation or standing of the other CNR faculty members, although, as we have seen, the contention was also the product of other concerns, such as the role of industry in the university and, to a lesser degree, biotechnology. It is likely, then, that the agreement would have been the subject of at least some conflict in any case.

In conclusion, it appears that UCB-N resulted in modest benefit and very little harm to PMB. A handful of faculty members benefited from temporary access to the proprietary data and advanced research technology made possible by Novartis. The majority of people in PMB benefited only to the extent that a significant amount of money ($60,000–$200,000 per annum) was made more easily available to them, in contrast to the more arduous route of governmental competitive grants. The graduate students who entered PMB between 1998 and 2002 benefited the most from the temporary increase in departmental financial support but appear to have experienced no other benefit from the agreement. It seems that students also suffered no obvious harm. Whether their socialization to more permissive university-industry relations represents a decline in their moral and ethical standards we will leave for another discussion.

Most PMB faculty viewed their experience with UCB-N as enhancing their academic freedom rather than restricting it, contrary to the expectations of opponents. Proponents of UCB-N tend to stress the relatively unfettered character of the research conducted using UCB-N funds, while critics point to the grant's built-in incentives to work on a circumscribed set of topics using certain equipment and datasets. It might be argued that both sides have a point, in that they view academic freedom quite differently.

TEN

The Impact and Significance of
UCB-N on UCB and CNR

I N ITS FIRST STATEMENT on academic freedom, the fledgling American
Association of University Professors (1990 [1915], 393) noted that "Academic freedom ... comprises three elements: freedom of inquiry and
research; freedom of teaching within the university or college; and freedom
of extra-mural utterance and action." In our inquiry we found no direct evidence of concerns about the first or second of these, but some concerns about
the third.

UCB's agreement with Novartis produced few or no effects on the freedom of faculty to engage in research and teaching. Moreover, there seem to
have been few direct negative repercussions for faculty and graduate students
who opposed the agreement. However, one CNR faculty member who publicly opposed the agreement claimed that his funding was reduced and that
he was pressured into moving into a smaller laboratory space because of his
opposition. While such actions are of great concern, there is little evidence that
faculty academic freedom more broadly was constrained in such palpable ways.

More significant for Berkeley and its future is the question of whether
UCB-N, and relations with industry more generally, transformed research
orientations in ways that threaten the academic freedom of certain faculty
members. This question was pertinent only for CNR, and not the whole university, as UCB-N has had no known direct effects on research outside CNR.
Many faculty who were opposed to the agreement argued that research critical
of conventional agriculture was increasingly being pushed to the margins and
not rewarded by the administration. In the opinion of these faculty members,

this marginalization and devaluation was happening because of the UC administration's increasing emphasis on entrepreneurialism and the concurrent strengthening of relations with industry. For these critics, the agreement with Novartis was the latest step in the restructuring of CNR along these lines.

A number of faculty, most of them in CNR but not in PMB, argue that UCB-N reflected a larger set of changes in the kinds of research that are rewarded at Berkeley. These critics charge, for example, that molecular reductionism is being rewarded at the expense of biological holism.[1] While it can be argued that biological holism was never very prominent at UCB, and while it is unclear whether it is in fact disappearing altogether as a field of study, some ESPM faculty claim that all the new positions in the biological sciences at UCB are in molecular studies. At the same time, a number of proponents of UCB-N claim that much of the opposition to it came from a small group of faculty who work in ecological or organism biology. Moreover, the divide was described as between those that work in biotechnologies and those who are critical of current technoscientific tendencies generally and genetic engineering specifically.

Some critics argue that research critical of the new technologies is increasingly being marginalized, while research that seeks to generate new technologies and forms of technoscience, such as genetic engineering, is embraced and promoted by UCB. A third division is between the "haves" and the "have-nots." The "haves" are practitioners of "big science" with expensive facilities and large extramural grants who tended to support the agreement, while the "have-nots" are faculty who conduct their scholarship with simple equipment and relatively small grants and who largely opposed the agreement.

The ecologists, the critics of biotechnology, the have-nots, and faculty who do research on alternative agriculture feel that the UC administration views them and their research as unimportant. They believe that Berkeley is increasingly promoting scholarship that generates large grants or gains the university public prestige, while work that is critical of the current social, economic, political, or scientific order, or generates little external funding, is being marginalized. A significant number of CNR faculty thus feel disenfranchised.

For some of these critics, particularly advocates of sustainable agriculture, the rise of biotechnology in agriculture is viewed as synonymous with increasing industry involvement in university research. While industrial involvement in the agricultural sciences is nothing new, the character of industrial involvement has clearly changed with the emergence of biotechnology (Kloppenburg 1988; Krimsky 2003). For example, the pesticide testing of the 1960s was more a service/extension function than a form of research. Faculty were paid by private firms to test specific pesticides. With biotechnology, the penetration of corporations into universities is greater. Biotechnology firms seek to engage university scientists in developing new technologies—in doing research on their behalf—rather than simply testing products. Many CNR critics of the Novartis agreement argue that industry involvement has skewed funding and stature toward research in biotechnology. They point out that there are natural affinities between the interests of industry and certain kinds of research. Faculty who work in the area of biotechnology are seen as having a lot in common with industry and thus as able to get industrial funding with greater ease. CNR is thus tilting toward molecular and genetic research, these critics claim, and away from research on the social and environmental implications of modern agriculture.

The treatment of the Gill Tract proposal and, particularly, the ensuing controversy over Quist and Chapela's article (see Chapter 4) reflect the conflict in CNR over academic diversity and freedom. Consider the Gill Tract, the largest continuous stretch of agricultural land in the east Bay Area. It is owned by the California Agricultural Experiment Station and is the former home of the Division of Biological Control. In 1997, at about the same time that UCB-N was under negotiation, a coalition of sustainable agriculture faculty at Berkeley and the Bay Area Coalition for Urban Agriculture proposed to make the Gill Tract into an urban agricultural research center. This, they argued, would enable UCB to fulfill its land grant mission within its urban setting.

The advocates of this project say that Rausser and the CNR administration never seriously considered the proposal. For many, both inside and outside UCB, the disparity in the treatment of the Gill Tract proposal and of UCB-N heightened awareness about the divergent directions of public research on agriculture. Given the choice of an alliance with the community or a deal with private industry, they argue, UCB chose industry. In June 2000 the urban agriculture project was rejected; the future of the Gill Tract remains undetermined, though the university ordered agricultural research to cease in 2003 as the administration developed plans to construct a dormitory, supermarket, and ball fields on the site.

The recently revised university guidelines on academic freedom theoretically allow faculty to be committed to "definite" points of view. The revised guidelines, which went into effect in September 2003, state, "Although competent scholarship requires the exercise of reason, this does not mean that faculty are unprofessional if they are committed to a definite point of view" (University of California 2003, APM-010). In practice, however, not all points of view are treated equally by the UCB administration. For a number of reasons, the UC system and UCB administration have committed themselves to certain areas of research over other areas, and to stronger ties with industry. While such commitments potentially increase the academic freedom of some, they may reduce the academic freedom of others. For many faculty, students, and activists a decreasing emphasis on agroecological sustainability is a notable threat to the diversity of views in academic research and teaching within the UC system.

COLLEGIALITY AND GOVERNANCE

Collegiality means, in part, the capacity to tolerate academic diversity. The tension between faculty members with different research orientations is perhaps most visible in the poor collegial relations typical throughout much of CNR. UCB-N did not create these tensions and differences, but it did throw them into sharp relief. Without question, it has also exacerbated them.

The poor collegial relations between departments in CNR can be attributed in part to the different perspectives and political commitments of the faculty. The departments and divisions of CNR are connected only loosely, and only insofar as they all focus on some dimension of food, agriculture, or the natural environment. The most visible divide is between faculty who work on sustainable agriculture and those who work on biotechnology. This divide is particularly evident in differences between PMB and ESPM. Both departments are interested in similar problems, but they understand and approach them from different perspectives. For example, two faculty members, one in ESPM and one in PMB, noted independently of each other that PMB faculty believe that the world can be engineered to be a better place. ESPM faculty, they agreed, tend to believe that ecologically informed systems are best and that humans should adapt their practices accordingly. While this characterization of the difference between the departments is almost certainly a radical oversimplification, it does illustrate significant differences in ideological commitments.

While several of the people we interviewed noted that collegiality improved during the time frame of the agreement, partly as a result of changes in the administration of CNR, most continued to see serious divides among the faculty in the college. When asked about collegiality, CNR faculty and graduate students said things like "the college is poisoned," "we have a fractured community," "there is little interaction," "relations are adversarial," "you can't trust people anymore," and there is "friction between graduate students." Members of the central administration largely concurred, remarking that the college is somewhat polarized. A number of ESPM graduate students also noted that UCB-N and the ensuing controversies have made cross-disciplinary discussions on biotechnology and sustainability nearly impossible.

In sum, relations among faculty in CNR continue to be a serious problem. Such a poor state of collegiality hinders the productive capacity of the college as a whole and the quality of education that it is able to provide, especially at the graduate level. In an era in which both social and environmental problems often require multidisciplinary efforts, disciplinary boundaries are becoming entrenched in CNR.

This is a separate issue from public concerns about UCB-N, as articulated in the popular press. As we have seen, the mainstream media tended to portray the agreement with Novartis as a Pandora's box or an instance of technology out of control. The public is interested in issues of food safety and environmental protection. Yet limiting industrial influence on universities so as to protect academic freedom does not necessarily address these concerns. Traditional expressions of academic freedom in the natural sciences tend to view any targeted research negatively, including applied research on safety and regulatory issues. Moreover, there appears to be little enthusiasm on campus for exchanging the yoke of an industrial patron for that of a government patron who might also seek applied results. At the same time, scientists who want to pursue an unfettered, independent research agenda need to convince potential patrons of the value of their research. Any patron with an agenda will fail to satisfy the ideals of pure science and academic freedom. Given this reality, the scholarly debate on campus may have matured in a way that does not resolve the apparent friction between the public and scientific experts.

Another way in which collegiality relates to academic freedom is through shared governance. Intensification of an already remarkable system of shared faculty and administrative governance within the UC system helped to model the system in its current form. Douglass (2002, 4) argues that one aspect of the university that makes it unique is its "tradition of shared-governance: the concept that faculty should share in the responsibility for guiding the operation and management of the University, while preserving the authority of the University's governing board, the Regents, to ultimately set policy." Faculty participation in shared governance has declined significantly, however, as administrators have come to operate more independently of the faculty body. In this context, Clark (1997, 38) argues that shared governance "only works when it is shared to the point where some academics sit in central councils and the rest of the academic staff feel they are appropriately represented, or where decision-making is extensively decentralized to deans and department heads and faculty sit close to these newly strengthened 'line managers,' or in various other complicated combinations of centralized and decentralized decision-making."

The key here is that not only was UCB-N negotiated in a way that made faculty feel excluded from negotiations but that there was a history and a context for this: Many had already experienced contract oversight committees composed of industry, government, and administrative staff that—at best—sidelined faculty concerns and interests. These faculty looked at the faculty seats on the UCB-N oversight committee filled by PMB scientists and had to wonder again whether the wider interests of the university and its faculty would be genuinely represented. One faculty member we interviewed said, "They're operating on our behalf; if they screw up it's our name that gets [tarnished]."

REPUTATION

The effect of UCB-N on the reputation of Berkeley is disputed. Both proponents and opponents of the agreement largely agree that the initial controversy generated negative public relations for the university, but proponents tend to argue that in the long run UCB-N has not negatively affected the reputation of the university. One argument in support of this position is that the good reputations of both Berkeley and Novartis served to overcome the initially negative PR. In other words, because Berkeley is an "honorable institution" known for its academic values, and because Novartis is a "reputable company," neither would enter into an agreement that would hurt UCB or its reputation.

One must also ask to what degree UCB-N was of interest to persons outside the UC system, whether in the biotechnology sector, in the field of agricultural sustainability, or in higher education. Contrary to the impression of many of those concerned with UCB-N, the controversy over the agreement hardly made the headlines at the national level. Moreover, we encountered very few people in the conventional agriculture sector in California who expressed concern; indeed, we found the majority we contacted had little more than peripheral knowledge about UCB-N. And external funding seems to have been minimally affected, if at all. Donald McQuade, vice chancellor of university relations at UCB, argued that UCB-N is "not something that comes up with respect to the campus's relationships with corporate and foundation donors."

By contrast, faculty and graduate student critics repeatedly described UCB-N as "tainting" or "tarnishing" Berkeley's reputation, while Peter Rosset, co-director of Food First, argued that UCB-N had left "a lasting smudge on the university's reputation." The primary concern of these critics was the legitimacy of the university and sustainability research done under its auspices. More conventionally, some faculty argued that research findings from Berkeley would no longer hold their traditional value in certain policymaking circles because they could no longer be considered "objective." More specifically, they claimed that the credibility of Berkeley, and the ability of its faculty to be impartial arbiters on issues of biotechnology, had been ruined. Faculty in CNR, and particularly within ESPM, were most outspoken on this point.

For those already critical of the trajectory of governance and research at UCB, the agreement with Novartis confirmed their doubts about the future of the university. For others, UCB-N was either viewed positively or, at worst, as a short-term smear on UCB's reputation, one that the university would quickly erase. It may be beside the point, however, whether the agreement hampered the ability of some CNR faculty to undertake objective research or participate in policymaking as independent experts. The perception itself has potentially important consequences. In considering future partnerships with industry, Berkeley administration and faculty clearly must be more attentive to the potential consequences for the university's reputation.

Public Mission

Given the details and disagreements of the UCB-N case, we now return to the question of institutional mission with fresh insight. The rise of universities was part and parcel of the rise of the nation-state and modern society; universities embraced a mission to distill research and teach national culture and natural science to the evolving citizenries of the nineteenth and twentieth centuries. David Harvey (1998, 199) puts it this way:

> In Europe it helped to solidify national cultures, gave "reason to the common life of a people," and fused "past tradition and future ambition into a unified field of culture." The university embodied an ideal. In the United States the mission was parallel. But here it was to deliver on a promise— to create tradition, found mythologies, and form a "republican" subject who could combine rationality and sentiment and exercise judgment within a system of consensual democratic governance. The university was where elite citizens went to be socialized and educated.

LGUs popularized this practice, their original mission, as set forth in the first Morrill Act, being "to teach agriculture, military tactics, and the mechanic arts as well as classical studies so that members of the working classes could obtain a liberal, practical education" (NASULGC 1995, 1). In part as a result of the intensification of rural-to-urban and international migration, combined with the rising complexity of national divisions of labor, race, and class structures, the democratization of education itself increased—often in line with the model developed by California progressives, which included three tiers of higher education.

Following World War II, in the context of cold war liberalism, civil rights struggles, environmentalism, and multiculturalism, the clarity and singularity of the idea of one universal science and one singular national culture increasingly came into question. The manifold constituencies served by the university came to engender what Kerr (1963) called the "multiversity," an institution with multiple—and at times contradictory—missions. More recently, the struggle for cultural inclusiveness and the diversity of science has been waged in the culture wars and the science wars that intensified at the end of the cold war and the rise of the information and risk society (Beck 1992; Castells 1991, 1996). Delanty (2001a, 158) optimistically addresses the uncertainty over the mission of contemporary public universities, given decreases in public funds and increases in academic entrepreneurialism and corporate grantsmanship, as follows:

> The implication of this for the university is that it will have to look beyond the nation for its cultural mission. Neither the capitalist-driven market nor postdisciplinary managerialism will provide the solution … to the challenges that technology poses. The solution resides in linking the challenge of technology with cultural discourses. Universities are among the few locations in society where these discourses intersect. As sites of social interconnectivity, they can contribute to the making of cosmopolitan forms of citizenship.

Like much of the debate regarding UCB-N, the impacts, if any, that the agreement has had on Berkeley's public mission vary significantly according to whom you ask. Proponents of the agreement, primarily PMB faculty and administrative officials at UCB, view close links with industry as largely compatible with the public mission. Critics, including faculty and graduate students working on social and environmental sustainability, and populist agricultural and environmental advocacy organizations, see research that is of direct benefit to private firms as antithetical to the public mission of universities. Research that benefits private firms is not necessarily of benefit to the public; for these critics, the public and private spheres should be kept distinct. We should note that a number of people on both sides of the debate asked whether UCB was still a public university or whether it had become a "publicly assisted" university. Proponents of UCB-N tended to have little problem with this transformation, while opponents of the agreement tended to see it as highly distressing.

Not only did PMB faculty who signed the agreement see little or no conflict between industry relations and the public mission of UCB, some of them claimed that collaboration with industry actually helped them fulfill the public mission. For one thing, it provided essential funding, given reductions in the state budget for education, for research that could benefit the people of California. Collaboration with Novartis also made the transfer of research into the public domain more efficient through increased patenting and licensing opportunities, they argued. For the UC administration's part, industry groups are understood as part of the public and as such are entitled to the same benefits as other constituents of the university.

Critics of UCB-N, who generally shared a more populist vision of the university's public mission, see things very differently. A number of these faculty members claimed that they were the "the people's researchers." They viewed the public mission of Berkeley as doing work that would benefit the public sphere and not necessarily the private sphere—doing work in the public interest that is not done by other groups. As one faculty member put it, "We're a public university and we have responsibilities to pay attention to people who can't pay for our attention." Thus, for many faculty who opposed the agreement, doing research for the benefit of private firms was antithetical to the public mission.

Other critics argued that the university's patrimony, invested in and paid for largely by the public, was being sold off for virtually nothing. A number of faculty commented that the $25 million from Novartis was a great bargain for the company and a loss for UCB, because, in their view, it gave Novartis the right to more than 135 years' worth of public investment.

Like these faculty members, most agricultural and environmental advocacy organizations also held a populist vision of the public mission. One member of an advocacy organization remarked, "We own [California's public universities], and they are not to be sold." Such organizations argue that all research produced in public institutions like Berkeley should be truly public (i.e., freely available outside the market). For them, UIR go against the public mission of Berkeley because they privatize research that rightfully belongs to the public.

Perhaps most important, and in line with criticisms of the trajectories of research in CNR and governance across the university, this argument suggests that prioritizing UIR could undermine the ability of universities to conduct research in the public interest that is not necessarily of interest to private firms. Criticizing not only UCB-N but also the management of the UC system, Peter Rosset of Food First remarked that UCB-N "demonstrates that there is no democratic mechanism between the taxpayers who are paying for a school of natural resources and the decisions that are made within that school of natural resources as to the ... emphases, or focus to take in the research activities." The privatization of UCB is of deep concern to organizations like

Food First; if public universities are privatized, they ask, who will do research on behalf of the public interest? As another representative of an agricultural/environmental organization put it, the public ownership of UCB is "absolutely critical to the survival of independent research and an effective educational system."

Clearly, the definition of the public mission is highly contested. The two sides of this debate obviously have different understandings of what is meant by "the public" and on the proper relationship between the public and private spheres. Proponents of UIR implicitly argue that the private mediation of the public benefits of research at state universities is broadly unproblematic, while, particularly in today's globalized world, critics explicitly question the benefits to California's public, writ large, of research sponsorship and patent licensing agreements with international corporations.

Land Grant Mission

Mirroring long-running debates over the meaning of the land grant mission, both at UCB and more generally, another point of contention in the UCB-N controversy was whether the agreement was appropriate at an LGU. The lines of disagreement here are similar to those in the debate over the meaning of UCB's public mission. PMB and the administration viewed close ties to industry as fully congruent with the land grant mission. Faculty critics of the agreement had mixed views on this question; there was less uniformity here than on the issue of public mission. Populist agricultural and environmental advocacy organizations argued most vehemently that industry links went against the land grant mission. They see biotechnology as generally serving the interests of large-scale agriculture and international agrifood conglomerates, and the interests of big agriculture do not fit the land grant ideal.

We learned through our interviews that most UCB faculty do not think much about the land grant mission as they go about their work of teaching and research. One PMB faculty member remarked, "around the department, most faculty would have a hard time explaining what the experiment station is." When asked, however, PMB faculty generally see collaboration with industry as fulfilling the land grant mission. They argued that collaborations with industry gave university researchers an additional path through which to transfer their research into the public domain. They also viewed UIR as a mechanism for turning basic research into useful applications consonant with the land grant mission. At the same time, they said, they would be doing the same work regardless of the land grant mission; thus it appears that the land grant mission has had little influence on most PMB faculty.

The Berkeley administration also tended to view UIR as consonant with the land grant mission, because they promote practical work connected with "real-world" needs. W. R. Gomes, vice president of DANR, commented, "The partnership of public and private can, in its best form, be an ideal way of making sure that important, current issues are addressed, and we are a mission-oriented division." Similarly, Paul Gray, the EVCP of Berkeley, remarked, "The industry connection and the funding of work here has a pronounced effect on making the work more relevant and less disconnected from real-world applications." For administrators, doing applied research fulfills the land grant mission. They view industry support as a highly efficient way to turn university research into "useful" products that will benefit the public.

Faculty critics of UCB-N took a more mixed vision of the relation between UIR and the land grant mission. Moderate critics argued that UCB-N does not represent Vice President Gomes's "best form" of public-private partnership, while more strident critics argued that the industrially funded and connected "real-world applications" stressed by Executive Vice Chancellor Gray generate more social and ecological problems than they solve. Along these lines, both moderate and strident critics tend to see collaborative arrangements like UCB-N as violating the university's land grant mission. They are aware that industrial influence is not new; several of them commented that California agriculture has been run by "big agriculture" for decades and that the land grant mission was taken over by agribusiness long ago. A number of UCB-N supporters made the same point. Commonly citing the "tomato harvester incident" (Friedland and Barton 1975), they argued that land grant funding, such as SAES funds, is also not necessarily independent of industry influence.[2]

In addition, some of the faculty critics of UCB-N did not believe that industry relations were necessarily in conflict with the land grant mission but saw them as one way to fulfill UCB's service obligation. They stressed, however, that careful governance was necessary to ensure that UIR were conducted appropriately—in a transparent manner and subject to some form of oversight. Many thought that transparency and oversight were lacking in the case of UCB-N.

It is important to keep in mind that the land grant mission is of little importance to most faculty and administrators at UCB. With the exception of some ESPM faculty, the land grant mission has little effect on the current faculty's research. In part, this is because the Davis and Riverside campuses generate most of the applied and commodity-group-focused land grant activities, whereas UCB has embraced basic research in agricultural economics, plant and microbial biology, nutrition, and the environmental sciences. As noted above, many faculty in CNR had little knowledge of the substance of the land grant mission. They more or less assumed that because they were doing work that they perceived to be in the public interest, they were fulfilling UCB's land grant mandate. For both administrators and faculty, then, the land grant mission is not of much relevance.

The degree to which a possible reworking of the land grant mission is necessary is suggested by the position of populist agricultural and environmental organizations, which argue that the land grant mission has long since ceased being a part of UCB. Representatives of these organizations point to what they see as a long history of UCB violations of that mission. Echoing long-standing populist critiques of the land grant system (e.g., Hightower 1973), they argue that the people of California lost UCB to agribusiness a long time ago, and that UCB-N is just one more example. As far as they're concerned, UCB-N benefited big agriculture at the expense of small farmers and alternative agriculture. This went against the land grant mission, which was originally intended to serve small farmers and the people of California. As Dave Henson, the director of the Occidental Arts and Ecology Center, commented, "the Land Grant was such a great idea, but it's only serving capital; it's not serving ecological health, which is one and the same as human health, and one and the same as sustainable economy." Agricultural and environmental activists are not the only ones who think that LGUs are increasingly controlled by private firms. This belief is also widespread in the farm community (Patrico 2001).

While some agricultural and environmental advocates agreed that industry ties provided UCB with much needed infusions of money, there was general agreement that increasingly close relations with corporate agricultural firms violate the land grant mission. Peter Rosset perhaps best sums up their argument.

> One direction is that the university ally [itself] with the private sector and get on the bandwagon of the trend of genetically modified crops and further industrialization of agriculture and appropriation of farming for private profit. The other direction was [that] the university would ally itself with the community surrounding it and lend its research and education support for alternative food systems and people farming at a smaller scale and in a nonindustrial way but playing an important role in community food security in the Bay Area.

CONCLUSION

Many of the issues at the center of UCB-N and the ensuing controversies are not new to Berkeley or to higher education. All the issues discussed in this chapter—academic freedom and diversity, collegiality, the reputation of UCB and its faculty, and the public and land grant missions—reflect the multiplicity of interpretations of the central principles of the university: creativity, autonomy, and diversity. For PMB faculty, UCB-N promoted and ensured each of these three principles. Non-PMB faculty see UCB-N as constraining their ability to conduct the kinds of research they see as important and valuable, which limits the creativity and diversity of research at UCB and leads instead to homogenization. Homogenization undermines not only CNR's reputation for cutting-edge research but also its ability to serve its many constituents, as certain kinds of research and forms of extension disappear.

UCB-N has raised a number of questions that relate to specific internal matters at UCB and in the UC system generally. Many of these questions are not restricted to the University of California, however, but reflect a larger debate about the future of public universities and higher education in general, to which we now turn.

ELEVEN

Rethinking the Role of Public and Land Grant Universities

ESPITE A FAIRLY BUMPY ride through the twentieth century—
encompassing two world wars, a global depression, a number of
recessions, and the cold war—institutions of public higher education retained their largely progressive character before starting to gradually
implode in the late 1960s. Starting with the emergence of the New Left, in connection with and following from the civil rights movement, the academy
served as a hotbed for every sort of liberal reform and radical program. At the
same time, though less visibly, universities remained intimately engaged with
conservative intellectual and cold war economic and military institutions.[1] In
this context, the stagflation of the 1970s, the recessions of the 1980s, and the
fiscal restructuring of the 1990s have generated changes in public funding of
the academy, which is as much about a cultural reaction to political correct-
ness, identity politics, and "leftist elitism" as anything else. Neoliberal economic and fiscal restructuring combined with neoconservative political and
cultural reform have thus contributed to a widely perceived crisis in the mission, purpose, and programs of public universities. This crisis is seen in controversies such as UCB-N and is inextricable from the intellectual debates that
dominate the culture wars and science wars (see, e.g., Gross and Levitt 1998;
Latour 2002).

Delanty (2001a) writes of four major changes associated with this crisis.
First, the dominance of the state in knowledge production begins to be balanced
(or countered) by new private and public-private research programs. Though
the state remains the primary financier of technoscientific development,

changes in patent law, information technologies, global trade in commodities, and global exchanges of cultural practices expand the arenas of knowledge production and application. Second, economic profitability, political influence, and everyday activities are more and more dependent on technoscientific, political, economic, and cultural knowledge. As such, the role of the university has had to shift as education—often of very specialized, as opposed to liberal and generalist, kinds of knowledge—becomes of greater importance to more people globally.

Third, mass education and mass movements have succeeded in altering the terrain and scope of university enrollments as lay and professional knowledge increasingly converge. Fourth, and finally, the democratization of knowledge has informed and resonated with the condition Beck (1992) first called "risk society" and later "reflexive modernization" (Beck, Giddens, and Lash 1994). Here, the combined growth in public understanding of the consequences of science and increasing numbers of science-based social, public health, and environmental crises has led to the rising contestation of knowledge and the primacy of expert-driven technoscience by a range of social movements. This has contributed to the partial delegitimation of traditional images of ivory tower academics and louder calls for academic, technological, and scientific accountability.

It is in this setting that Delanty addresses what others have called entrepreneurial science, or "academic capitalism," in the context of the "new managerialism" (Clarke and Newman 1997; Etzkowitz 2003; Etzkowitz and Leydesdorff 1997; Slaughter and Leslie 1997). In the same context Derek Bok (2003) has written on the commercialization of higher education. What this means is that, on the one hand, there has been a series of state-level fiscal shocks to public universities, and, on the other hand, there has been a society-wide restructuring of the public's relation to knowledge, science, and education. On another axis, while state-level support for general educational funding (teaching and administrative support) has fallen, federal support for big science has increased. Significantly, however, Delanty suggests that the current debate over the status of the university in an era of neoliberal fiscal constraints and economic and cultural globalization encompasses far more than academic capitalism and managerialism.

From the perspective of Bill Readings (1996), the twentieth-century university struggled to hold together the Kantian commitment to pure science and the Humboldtian commitment to public citizenship. During the cold war, widely shared economic commitments to technical and military efficiency, alongside social commitments to liberal civic reform, reduced tensions between these two tendencies—at least within the academy. In the context of the serial rise of the New Left and New Right, however, these tensions have reemerged with new force. From the perspective of the mid-1990s, six broad claims were made within the public discourse over academic restructuring. Advocates have suggested that the university needs to return, or develop a new orientation, to:

1. the inculcation of traditional liberal values (e.g., Bennett 1992; Bloom 1987);
2. critically evaluating contemporary social and cultural problems (e.g., Fuller 2000; Krimsky 2003);
3. the pursuit of (basic scientific and cultural) knowledge for knowledge's sake (e.g., Committee to Review Swedish Research Policy 1998);
4. training students for the (now global) job market (e.g., Thurow 1999);
5. providing innovative technology for industry (e.g., Clark 1998; Etzkowitz and Leydesdorff 1997); and
6. serving as an engine of regional and state economic growth (e.g., Nelson 1993; Varga 1998).

These six agendas fit the Kantian-Humboldtian divide, as the first three focus on the reproduction of scientific culture and the last three on socioeconomic development issues. What is clear, however, is that holding on to the idea of a singular, universal mission for higher education has become quite difficult.

In an article called "The University in the Knowledge Society," Delanty (2001b) reframes Readings's analysis into four contemporary approaches to the university's role in contemporary society: (1) the *neoliberal* critique of the decline of the Western canon; (2) the *postmodern thesis* of the end of a singular, universal culture and therefore of a single universe of knowledge; (3) the *reflexivity thesis* that suggests the university is in crisis—and that a new mode of knowledge production and consumption is being generated—as relations between the university and its economic, political, and citizen clients evolve; and (4) the *globalization thesis,* which argues that the university needs to be (or has already been) remade in the face of its new, instrumental associations with global capitalism and commodifiable information technologies.

Whereas Readings saw the university in ruins, Delanty argues more coherently that the university needs to re-vision its role in the knowledge society. Much like Bok, Delanty sees universities moving down all of these paths largely by an uncoordinated process of crisis management. The uncertainty of the combination of the culture wars, the science wars, and fiscal crisis management leads Delanty, Bok and others to believe it is a *lack* of a formal, public discussion about the nature and future of the university that has generated such extraordinary unintended consequences. The controversy surrounding UCB-N has been approached by people motivated by the different kinds of assumptions reviewed above. Whether they focus on the cultural, rational, political, scientific, or economic grounds upon which the university—and the historical agricultural college within it—are believed to properly stand, participants in the controversy have been presented with no opportunity for real dialogue.

The major advantage of analyses such as those of Readings and Delanty is that they see universities as necessarily tension-filled places: The diverse roles universities play and the diverse purposes they serve verge on the incommensurable. For them, Kerr's multiversity is contradictory and multifunctional, though contradiction and diversity of function are not necessarily to be avoided. Contemporary society manifests the same condition. If this is correct, and we believe it is, then in order to remain socially and scientifically viable universities must accept that they are the locale wherein these incommensurabilities meet, and where debate must be fostered rather than foreclosed. The democratization of the institution is fundamentally necessary for its future stability and survival—particularly if autonomy, creativity, and diversity (however defined) are to remain the institution's core values. Yet present trends appear to be toward fiscal, research, and teaching efficiency at the cost of democratization.

In our interviews, one of the most common criticisms of UCB-N was the ways that the *process* of its negotiation—and subsequent exchanges between administrators and faculty—illustrated a decline in faculty participation in the governance of the university. At UCB and across the UC system, existing structures of democratic governance need to be revitalized such that a regular, institutionalized discussion of the missions, responsibilities, and directions of the university are once again made central to an ongoing debate. At other public and land grant universities, greater shared governance needs to be instituted. The alternative is further bureaucratization, accompanied by greater managerial control and eventual decline into mediocrity.

VISIONS OF THE UNIVERSITY

The various documents we reviewed and the interviews we conducted reveal profound disagreement about the role of the university. While others might put it in somewhat different terms, we see three overlapping yet distinct models of the university implicit in both the documents and the interviews. We call the three models (1) universities as engines of growth, (2) universities as founts of societal betterment, and (3) universities as knowledge-generating institutions.[2] Few people subscribe fully to any one of these models. They are best seen as Weberian (1947) ideal types rather than as the actual views of real groups of people. Up to a point, the first two represent variants of the progressive and populist strains of the twentieth-century nationalist university, while the third distills the essence of the globalizing yet imploding twenty-first-century institution. The key is that the models serve to put the matter in clear relief, so let us elaborate further on each of them.

The University as an Engine of Growth

Historically, growth was not problematic in the behavioral and economic sciences; in general, it was viewed as either natural (e.g., Malthus) or serendipitous (Schumpeter). With the elaboration of the modernization project in the second half of the twentieth century, attention was directed toward the factors that fostered economic growth: culture, human capital, social organization, and technology. Culturally, individualism, inquisitiveness, and innovation were seen as supporting economic growth. It was argued that enhancements in human capital, both in general education and in specific occupational (and even civic) skills, contributed to the expansion of the economy. Particular forms of social and economic organization—democracy, free markets, trade, urbanization, the LGU—were viewed as fostering economic growth. And new forms of technology—whether energetic, physical, chemical, biological, or informational—gave rise to new goods and services in an expanded economy (and stabilized polity). The role of the university in the growth process derives from its relationship to each of these four factors.

The recent conceptualization of universities as "engines of growth" or "growth machines" was one of the outcomes of the kinds of growth just described. Initially it was an answer to an empirical question: How have the Four Tigers and the newly industrialized countries been able to do what they have done? Eventually, in the face of intensifying global markets and competitiveness, the concept evolved into policy efforts to identify new kinds of steering mechanisms appropriate to producing economic growth in a state, region, or country. At the national level, over the past quarter-century, the primary mechanisms have been associated with neoliberal de/re-regulation, the quasi-privatization of "voluntary" governance, and neoconservative forms of social discipline.

The downstream operationalization of the concept of growth has become contested terrain, where different social actors competed for social resources on the basis of claims about which factors produced the greatest economic growth. Four groups of claimants were particularly prominent in this contestation. A group of economists (e.g., Roemer 1970) emphasized the developmental impacts of trade and export-led development. Educators and labor economists (e.g., Roemer 1989) emphasized the growth that followed from upgrading human capital. Sociologists and urbanists (e.g., Molotch 1976; Scott and Storper 1986; Storper and Salais 1997) wrote extensively about the city as a growth machine. And, finally, advocates of different technologies (plastics, telecommunications) claimed that resources invested in their particular industries would produce the greatest return in economic growth.

During the 1980s and 1990s, biotechnology and information technology were viewed as the growth engines of the coming decades. Chemical companies (the growth industry of the late nineteenth century) built and bought significant biotechnological capabilities. Universities established biotechnology research parks where professors could undertake work on for-profit ventures (Buttel 1986; Isserman 2000). Large clusters of biotechnological R&D were created in the metropolitan areas of Boston, St. Louis, and San Francisco. Between 1980 and 2000 hundreds of biotechnology venture capital firms started up; although many paid good salaries and wages, few made profits or paid a return on investment.

Especially as the rate of growth in state financial support has decreased, universities have competed more openly for the "engine of growth" banner. Richard Levin (2001), the president of Yale University, has perhaps articulated this view in greatest detail: At least in the United States, universities are the principal source of new scientific discovery and thus of technological advance and economic innovation, and universities foster critical inquiry and creative leadership for industry and commerce. In an attempt to bolster its standing with the state legislature, the University of Wisconsin at Madison commissioned a study in 2002 on the impact of the university on the state's economy (Northstar Economics, Inc. 2002); the chancellor of the university used the results to argue against a reduction in funding that would hamper the university's ability to create high-paying jobs through the establishment of research parks. In 2003 the eight research universities in the Boston area compiled a report on the economic impact of the universities on the region (Appleseed, Inc. 2003). Many other universities have participated and continue to participate in similar efforts.

Geoffrey Owen (2003), the dean of biological sciences at UCB, put it most straightforwardly: "the state's education system has been the engine of economic growth in this state for decades." UC also employed the tactic used by the University of Wisconsin and the Boston-area universities, and commissioned a study of the growth impacts of the university on the regional economy. *California's Future: It Starts Here* is an independent report on the breadth and depth of the university's contribution to California's economic growth, health, and community resources. Among its findings are that UC's impact on the state economy exceeded $14 billion in 2002 (ICF Consulting 2003).

The importance of biotechnology, and strong institutional commitments to technology transfer more specifically, had real political resonance. Shortly before he was recalled, Governor Davis suggested that "A large part of California's future is going to be in the rapidly growing life science industry, which is not only an engine for economic growth, but holds the key to alleviating vast suffering and improving the health and well-being of literally every person in the world. This is why life sciences is [*sic*] going to be a key focus of our overall economic growth strategy" (quoted in *Sacramento Business Journal* 2003).

We noted in Chapter 7 that early newspaper coverage of the UCB-N controversy was presented in an economic frame, and that economic aspects were the third-most common theme in the news coverage. The economic theme was most often used in relation either to the economic needs of the university or to the economic benefits that the Novartis Corporation would gain from the agreement. It is notable in this context that the argument that the agreement would drive economic growth in the Bay Area or the state did not appear in press coverage. This may have something to do with Novartis's home being in Switzerland, rather than California, and appears to play into the hands of populist critics of the agreement.

Although the news coverage did not emphasize the possibility that biotechnology would be an engine of growth for the Bay Area, several university administrators and faculty did articulate this view. One noted: "More than ever, the university is seen as an agent of economic development, and there's no place where that's truer than in the Bay Area." That person noted that 30 percent of the world's biotechnology firms are in California and 30 percent of those were started by UC faculty. Another administrator suggested that upper administration was emphasizing the view that the agreement made it possible for faculty to make the kinds of discoveries that have commercial value, which in turn (at least indirectly) promotes economic growth. Yet another administrator argued that the agreement "could also be a benefit to the California economy. Even though most of the PMB faculty weren't doing applied research, the faculty were aware that there could be economic benefits from their work."

One faculty member acknowledged the "engine of growth" spin that the administration was putting on the agreement, but argued that it was premised on the assumption that an "industry-driven trickle-down approach" would result in economic development. Another faculty member suggested that the UC's realization of its role as an engine of growth acted as a model system for other countries. "[T]hey all look at it as ... this is the source of economic development. All the new companies are being started up and all the industry, in the biotech industry, in the computer industry, all start at the university."

Empirical research, however, suggests that (1) regional growth requires a combination of factors, including vital social, economic, and technical infrastructures at the local level (Washburn 2005), and (2) most biotechnology centers have added only marginally to regional economic prosperity (Cortwright and Mayer 2002). In a similar vein, we interviewed a public official involved in the controversy who felt that the Price and Goldman (2002) study uncritically accepted the assumptions of the "engine of growth" model as it applied to the agreement. The official believed that the model, in order to work, should be turned the other way around, to make "sure that the private sector is accountable to the public."

The University as a Source of Societal Betterment

Those who adhere to the notion that the university is a source of societal betterment take a different approach—one that is fundamentally transformative. They see the university not as a means of promoting growth in a predetermined direction but as a means of remaking society based on shared changing and emergent ideals. Proponents of this model have a more or less well-defined program for the refashioning of society.

First, they argue that research should generate knowledge that is a public good. Given the agrarian origins of the nation, it should be no surprise that at LGUs, farmers were the initial beneficiaries of such public goods. Colleges of agriculture poured forth a wide range of farm innovations, mainly of a biological sort, with the express purpose of helping farmers to raise their incomes and levels of living. Of particular note is that the bulk of these innovations were in the form of improved seeds or cultural practices. Until recently, neither was subject to IP laws. Moreover, the Cooperative Extension Service, an institutional innovation to which federal, state, and local governments contribute, was established to further the rapid spread of technical change. Programs in home economics and 4-H (for youth) ensured that farmwomen and children also received the benefits of public research.

Second, and following from the first point, proponents of the societal betterment model argue that knowledge should flow freely both within the university and to various more or less well-defined clientele groups of the university. If knowledge is power, and this is a democratic society, then knowledge and power should be available to everyone, irrespective of their wealth or income.

Finally, proponents of this model see education as preparation for citizenship in a democratic society. Indeed, the Morrill Act of 1862 establishing the LGUs spoke clearly of promoting "the liberal and practical education of the industrial classes." Higher education in a democratic society is not to be reserved for the rich; it is to be made available to everyone, whatever their class position. Better-educated citizens, it is claimed, participate more fully in civic life, making better decisions not only about their home and work but about the kind of government they want. They also further the American ideal of equal opportunity.

The University as a Generator of New Knowledge

Others interviewed adhered more closely to the final model we discuss here, that of the university as a generator of new knowledge. Proponents of this model note its several distinguishing features. First, knowledge is desirable for its own sake. The university is seen as a great cathedral of knowledge in which faculty are the masons and stonecutters who build the edifice. Proponents of this model argue that the immediate usefulness or utility of knowledge is of little importance. What counts is that it is generated. The pursuit of truth unhampered by the cares of the workaday world is central to this view.

It follows from this model that education is a matter of learning facts about the world and methods for exploring them, and of debates about society and culture and the critical traditions brought to bear on such questions. On the one hand, students are to take classes in order to learn the facts, and the history of those facts, to participate in canned laboratory experiments to demonstrate truths revealed by previous generations of researchers and to learn the terrain across which subsequent experimentation must move. On the other hand, students are to take classes in order to learn the material and semiotic practices that underlie contemporary societies and to engage in social and interpretive research exercises to hone their own critical faculties. Most important, however, the overarching goal of liberal education is to foster a synthetic and generative appreciation of the scope and complexity of knowledge in students' minds and lives.

For proponents of this model, too, free circulation of knowledge is essential. But the aim of that free circulation is quite different from that of the societal betterment model. Instead, they argue, knowledge must circulate freely so as to support the goals of the academy itself in its eternal quest for truth. This is the vision of the Ivory Tower or the City on the Hill.

Consequences

The reader will have little difficulty recognizing strong elements of utilitarianism, pragmatism, and positivism, respectively, in the above scenarios. Indeed, this is not entirely accidental; initially, we considered describing them in precisely those terms. On further reflection, however, it became clear that the three models were not entirely congruent with the three schools of philosophy. One important reason for the distinction is that the three models were formed in the rough-and-tumble world of practice rather than in scholarly academic debate. They make few claims about the ontological nature of the world, and only a few more about the epistemological nature of knowledge. Or, to put the matter differently, the practice of philosophy is quite different from the practice of those who create, modify, and transform universities. Most philosophers worry little about budgets, funding, licensing, building construction, police and fire protection, legislative mandates, or public demands in developing philosophical positions. They *do* worry about logical consistency, convincing others, and producing an adequate set of claims.

In contrast, those who run universities—and especially large research universities—are faced with an endless set of emergencies, crises, reorganizations, and the task of running what are and must be political institutions. They have little time, perhaps too little, to reflect on the consistencies or lack thereof in their views and actions. Theirs is reconstructed logic, rather than logic in use (Kaplan 1964). Nevertheless, a clear pattern emerges: The vision of the university as an engine of growth is now the dominant view. Lamentably, this has occurred with little or no real debate among faculty, students, or external constituencies.

We would do well to remember both parts of Eisenhower's warning, quoted in the epigraph to this volume: He warned us about the need to avoid both the domination of university faculty by money and the capture of public policy by scientific-technological elites. We now face the possibility of such scientific elites dominated by money *and* in charge of public policy. That does not bode well for the critical inquiry necessary to sustain democracy. The problem surfaces in the complex debates about conflicts of interest for faculty, administrators, and universities themselves.

GROWTH MODELS AND
CONFLICTS OF INTEREST

A worrisome consequence of the model of the university embodied in UCB-N was a perceived conflict of interest (COI) widely reported in the popular and scientific press. In a flurry of media attention, UCB and LGUs more generally were accused of selling their integrity. Press reports illuminated the tension between a public university's industry ties and its broader responsibilities to the public. UC's written guidelines advise that in "pursuing relationships with industry, the University must keep the public trust and maintain institutional independence and integrity to permit faculty and students to pursue learning and research freely" (University of California, Office of the President 1989, 1). As the discussion above on academic freedom and diversity indicates, it is open to debate whether the university fulfilled this obligation in its agreement with Novartis.

In fact, UC administrators acknowledged that controls on research integrity deserved a second look. For example, the president of UC at the time pointed out that agreements with industry "contain complex IP matters and may involve conflict-of-interest issues that do not typically arise in government-sponsored contracts and grants." The university's goals are "not simply to generate royalty revenue and stimulate economic growth but to create relationships with industry that will help faculty in pursuing their own research and in training graduate students." Individual conflicts of interest are "not easy issues to resolve," and "ensuring the integrity of university research" calls for "open discussion within the academic community" (Atkinson 1999, 47).

The idea that objectivity may be compromised by self-interest is hardly new. It is typically addressed through attention to professional ethics (or codes of conduct) and COI policies. The public and well-tempered policies of UC provide a useful illustration for research universities in general. Individual conduct is spelled out in considerable detail in UC's code of conduct. The university's COI policies conform to state law aimed at protecting the public interest. They define a conflict of interest as "a situation in which an employee has the opportunity to influence a University decision that could lead to financial or other personal advantage, or that involves other conflicting official obligations" (University of California, Office of the President 1989). While the formation of COI policy in the 1980s and 1990s focused on entrepreneurial faculty and their external financial interests, UCB-N raised issues of a different sort. In this case, it is the *institution's* potential for COI relative to funds it receives that is at issue.

UC's existing COI policy and procedures concentrate on the financial implications of licensing agreements and the governance of on-campus research. Yet the financial return from agricultural biotechnology is not determined solely by the terms of IPR agreements. The financial interests of individual researchers, academic units, UCB, and UC as a whole are also linked to governmental regulatory decisions, the documented results of field trials, findings on environmental consequences, findings on alternative approaches to agriculture, findings regarding agricultural economics, and the state of the public dialogue on biotechnology. The degree to which individuals affiliated with UC inform or influence any of these activities, either directly or through other organizations, could constitute a potential conflict of interest. For example, a faculty member who uses his or her influence to help persuade the Environmental Protection Agency that a new plant variety will have no harmful effect on the environment could yield financial benefit at public expense under current IPR agreements. In the same fashion, if the institution reduces its support of environmental research while holding a financial stake in the success of agricultural biotechnology, the same adverse trade-off could occur. These scenarios illustrate conflict deriving from a duality of purpose or a conflict of mission. Perceived COI endangers the credibility of fair and transparent agricultural, regulatory, and environmental research.

Standard attention to conflicts of interest and codes of conduct may well fall into what Power (1997, 123) refers to as "shallow rituals of verification." In effect these procedures do not begin to address the depth and breadth of the issues at hand. The debates over UCB-N suggest that the boundaries of current university COI policy are unrealistically narrow in several respects. Ideally, research related to the possible consequences (including risk assessment and regulatory science) of agricultural biotechnology should be secure and independent of both individual and institutional conflicts of interest. This is no simple task. Scientists may influence the regulatory regime through multiple channels, including (1) engaging in scientific dialogue (e.g., scientific conferences); (2) contributing to the scientific literature through submissions or through the peer review or editorial process; (3) managing and assessing IP; (4) engaging in public communications, including interactions with the media; (5) participating in boundary work and supporting boundary organizations; (6) conducting or choosing not to conduct research specifically for regulatory purposes; (7) serving directly on scientific advisory panels or in similar roles; and (8) consulting with industry on related matters.[3]

Controversy over UCB-N highlights a growing uncertainty about individual versus institutional responsibilities. The landscape surrounding the agreement suggests that public institutions need to cast a broader net in order to capture the full range of COI issues. Even with growing attention to institutional conflicts of interest, the agreement brings existing COI policies into question for at least four reasons. Current procedures generally (1) do not capture the range of actor influence in highly regulated markets, (2) poorly integrate potential financial and nonfinancial conflicts, (3) fail to acknowledge the potential detrimental influence of institutional self-preservation, and (4) do not capture the influence of an extensive web of boundary organizations. Moreover, the university's own accountability appears to lag behind its growing role in managing industry relationships and IPR. For example, UC guidelines assert:

> It has long been recognized that the only truly effective safeguard against conflicts of interest situations is the integrity of the faculty and staff. A codification of the complex ethical questions involved, even if possible, would be unduly restrictive. At the same time, even the most alert and conscientious person may at times be in doubt concerning the propriety of certain actions or relationships. Whenever such doubt arises, the University expects the individual involved to consult with the Office of the Chancellor, or the Chancellor's designated representative, before making a decision. (University of California, Office of the President 1989)

In this passage, UC policy refers to the "integrity of the faculty and staff," yet mandatory assignment of inventions matched with the terms of UCB-N could circumvent the faculty's control of IP of their own creation. Faculty members have an obligation to disclose patentable inventions, and the university has a contractual obligation to report these to an industrial sponsor. Under UC's patent policy (effective October 1, 1997), the president is responsible for managing "IPR for the public benefit" in consultation with the Technology Transfer Advisory Committee, chaired by the senior vice president for business and finance, who is responsible for implementation of this policy and may grant exemptions from it. However, his or her charge under the policy is limited by "overriding obligations to other parties."

The question of control was also raised in a laboratory study at the University of Wisconsin. Kleinman provides a detailed case study (conducted in 1995) of a laboratory led by Jo Handelsman in the Department of Plant Pathology, University of Wisconsin, Madison. Within this environment, he noted that for Handelsman, "the reality of running a university biology lab does not allow ... the luxury of separating 'the science' from matters of patenting, funding, and administration that play important parts in her professional life" (Kleinman 2003, 158). While still optimistic that such externalities can be isolated from "the science," Kleinman concludes with a continuing concern for university research as an engine of growth and doubts whether the pursuit of IPR actually enhances scientists' control over the fruits of their research.

Universities also have a financial interest in increasing research dollars for respectability or self-preservation. With this in mind, a *lack of commitment* to biotechnology, relative to other approaches with potential advantages to agriculture and the environment, may also generate a perceived conflict of interest. Even though the management of *non de minimis* ownership by faculty is well advanced, institutional COI remains a largely open topic. However, institutions generally ask individuals to completely disclose any *non de minimus* financial interests.[4] Institutional conflicts of interest related to biotechnology attracted more attention after the agreement with Novartis, in part because of a death at the University of Pennsylvania in 1999 (Barnes and Florencio 2002). Attempts to engineer viruses for medical uses (coined "virotherapy") include injection of genetically engineered viruses into human subjects. This experimental technique was apparently responsible for a death during clinical trials (Nettelbeck and Curiel 2003). In this case it was found that researchers and the institution had equity interests in the company (Genovo) that would have profited from the experimental gene therapy. Since the incident, the Association of American Universities (2001) and the U.S. General Accounting Office (2001) have expanded the COI discussion to include institutional finances.

In addition, reference to the university in this context is somewhat misleading. A wide variety of boundary organizations have emerged to mediate between public and private institutions. For example, to prepare a report for policymakers on genetically modified organisms, the General Accounting Office worked with the National Academy of Science's National Research Council and received advice from a long list of boundary organizations, including the Association for Analytical Communities International (formerly the Association of Official Analytical Chemists), the American Oil Chemists' Society, the American Association of Cereal Chemists, the Center for Science in the Public Interest, the Union of Concerned Scientists, the Biotechnology Center of the University of Illinois, the Health Sciences Center of Tulane University, the Consumer Federation of America, the Council for Agricultural Science and Technology, the Pew Initiative of Biotechnology, the Biotechnology Industry Organization, the Institute of Food Technologists, and the United Nations Food and Agriculture Organization (U.S. General Accounting Office 2002). A review of the same question produced by the California Council on Science and Technology (itself a boundary organization) included input from the California Institute of Food and Agricultural Research, the Competitive

Enterprise Institute, Greenpeace, the Grocery Manufacturers of America, the Hoover Institute, the National Food Processors Association, the Organic Consumers Association, and the Union of Concerned Scientists (California Council on Science and Technology 2002). A university's sphere of influence includes many of these entities, each of which can influence activities in the private sector. Therefore, institutional COI policy is inappropriately limited if it does not look beyond the university to incorporate language addressing relations with boundary organizations.

Recent attention to the financial holdings of researchers also overshadows other motivations for scientific misconduct. Other interests among researchers include the personal satisfaction of solving one of nature's difficult riddles, and professional positions, recognition, or public esteem. Financial conflicts are not needed for questionable behavior on the part of researchers or university leadership. While most policies address financial incentives derived from individual and institutional commercial interests, nonfinancial conflicts may have the same if not greater detrimental influence.[5] Individuals, especially in highly specialized fields, and institutions with customized assets may have a sufficient potential for loss from outside commercial ties without financial COI. Numerous governmental, legal, and regulatory decisions can precede the investment of significant private capital. At this early stage, when the agenda is still fluid, academic scientists inform the process through many distinct avenues. Early on, few researchers, if any, would have financial ties to industry relative to a scientific innovation. Therefore, financial disclosures in the critical pre-lock-in period are unlikely to provide much insight on potential COI.

The management of conflicts of interest must address both real and perceived conflicts. Even if opponents of biotechnology are shown to be misguided alarmists at a future date, the perception of a conflict exists and should be managed forthrightly. In this light, individual and institutional decisions within a broader scope could be brought under COI review. It is also routine to acknowledge that the establishment of new controls does not, in and of itself, imply any past wrongdoing.

In sum, our interviews revealed three somewhat incommensurable models of the university. There is little doubt that the growth model is currently dominant in most major research universities. With the dominance of that model, conflicts of interest are far more likely than they have been in the past. While universities have gone a long way toward developing conflict of interest approaches for individual faculty members, they have experienced much greater difficulty in applying such models to administrators, individual units, or universities as a whole.

Clearly, eliminating the growth model of universities entirely is both unlikely and inappropriate. Universities are faced with the continuing challenge of serving different elements of society in a wide range of only partially compatible ways. We cannot retreat to an idealized past, nor can we blindly move forward to an unimagined future. In the following chapter, we make some modest recommendations for rethinking and reenacting the university.

TWELVE

Constructing the Future

Re-visioning Universities

I N THE INTRODUCTION we noted that core principles are at the center of an ethical framework. In this concluding chapter we draw conclusions and offer recommendations with an eye to ethical discernment. UCB-N acted as lightning rod for numerous inarticulate concerns about the role and purpose of the university. Both supporters and detractors claimed to take the high road with respect to what we argue are the three core principles of the university: creativity, autonomy, and diversity. In some sense, everyone is right. But what has been obscured by the day-to-day business of the university—the teaching, research, and service obligations of faculty, the myriad requirements of managing a large and dynamic institution, the financial exigencies—is attention to the broader goals and objectives of the university. As Charles Vest (2005, 80) argues, "America knows better than any other nation how to do research, but we have lost our common understanding of why do research." These issues have received scant attention, as ad hoc, short-term decision making has consumed faculty and administrators alike. This is true at UCB, but it is also true at nearly every public research university in the United States.

First, although there is widespread agreement on the need for faculty autonomy at the university, the various (usually implicit) definitions of autonomy are vague and perhaps antithetical. Of what does autonomy consist? What do we mean by academic freedom? How ought we to try to reconcile the disparate goals of diverse fields of endeavor? Is academic freedom merely the freedom of individual researchers to pursue their interests, or must we begin to consider the concept as it applies to departments, centers, and university administration as well?

Second, there is also widespread agreement on the need to support and even bolster creativity among the faculty. But here, too, the definitions are vague. Can universities do this in a manner that fosters rewards for some without penalizing others? Can they develop mechanisms for enhancing creativity that are inclusive rather than exclusive? Can they better define what kinds of grants and contracts best serve to enhance the creativity of some faculty without limiting the creativity of others? Can the risks of creativity outside the box be acknowledged? Accommodated? Embraced?

Third, there is widespread support for diversity. But diversity means different things to different people. Here, too, universities face what appear to be trade-offs between enhanced intellectual diversity within a given field of endeavor and across fields of endeavor. Is there really a trade-off? If so, how might universities ensure that it is equitable? If not, how do administrators and faculty determine whether to pursue greater depth in a given field of scholarship versus supporting multiple fields of scholarship?

Fourth, the academic community must ask whether the three principles of autonomy, creativity, and diversity are in fact still at the core of higher education. If so, how does autonomy accommodate democracy (and fiscal crises)? How does creativity work across research, teaching, and service? How does diversity relate to difference? Do the principles need to be modified? Replaced? Reordered?

Finally, decision makers within the university must ask again how the institution should distribute goods in order to achieve its goals. What mix of markets, need, and dessert is appropriate for teaching, research, and service? How can the university reconcile these disparate approaches to the distribution of goods in a way that achieves the common good? What kinds of consensus, or even dissensus, exists with respect to defining the breadth, depth, and scope of the common good?

These are difficult questions, and they have no simple answers. But failing to address them is likely to be far more damaging to the future of research universities than continuing to ignore them and proceeding virtually blindly into the future. Clearly the future will not be a mere repetition of the past. But it will be molded, perhaps not just as some would like, by decisions made today not only on university campuses but by state and national legislative bodies and even the courts. Academia needs to examine its policies and procedures and see whether they fit together in ways that are meaningful and serve the common good.

Ethics will play a role in a reasoned response to these fundamental queries. Within university-industry collaborations and during their construction, there are ethical choices to be made. Addressing UCB-N, Yost (2003, 41) recommended that the "basic academic values for knowledge creation should go hand in hand with basic ethical principles when it comes to university-industry linkages for commercializing university technology." The UCB-N case study does not depict key participants laboring over ethical dilemmas, however. For proponents, the ethical standards embedded within the Office of Technology Licensing often seemed sufficient. For opponents, widely diverse institutional, environmental, and social ethics trumped the legal and political economic concerns of OTL. If ethical discernment should have been at the center of discussion, was it, or how was it, kept at bay?

To address the ethical considerations, one must compare the Berkeley experiment with university-industry collaborations deemed more successful. Adding counterpoints to UCB-N also helps us understand decision making within a realistic range of uncertainty. In like fashion, the transfer of knowledge needs to be shown in its multiple facets to capture the range of available choices—up and down different ethical and institutional streams. From this foundation, rhetoric teetering on a basic/applied dichotomy may be added to the analysis. It is on the nebulous line between "basic" and "applied" research that the ethical door of science appears to swing open or closed. Finally, we need to look at the distinction between individual and institutional conduct from both sides if we are to explore the ethical terrain thoroughly.

A CASE IN CONTRAST:
WOODWARD RESEARCH INSTITUTE

The history of science abounds with theoretical and practical work creatively intertwined, perhaps inseparable. This synergy often draws from vibrant university-industry linkages and can lead to significant results. The UCB-N case study may be placed in a different light when compared to similar events with far more affirming outcomes. A case in point is the Woodward Research Institute, established in 1963 by Novartis's precursor, Ciba Corporation. The institute and its namesake, Robert Burns Woodward (1917–79), illuminate the promise of university-industry collaborations. A native of Boston, Woodward quickly became an outstanding scientist. He began studies at MIT at the age of sixteen, earned a PhD there at twenty, was a full Harvard professor at thirty-three, joined the National Academy of Sciences at thirty-six, and won the Nobel Prize for Chemistry at the age of forty-eight.

Even with these credentials, one cannot properly tally Woodward's scientific contributions without mentioning his ties to industry. A World War II shortage of quinine (with applications in the treatment of malaria and photography) prompted early funding from the Polaroid Corporation. In publishing his pathbreaking synthesis of quinine in 1944, Woodward identified himself in the *Journal of the American Chemical Society* as an "instructor in chemistry at Harvard University and chemical consultant to Polaroid Corporation" (Morris and Bowden 2001, 58). When, as a Harvard professor, he reported the total synthesis of steroids in 1952, Woodward acknowledged contributions from Eli Lilly, Merck, Monsanto, the Research Corporation, and the U.S. Public Health Service (Morris and Bowden 2001, 114). Woodward made theoretical discoveries in organic chemistry at the same time that he served industrial interests. These contributions, among them the Woodward-Hoffmann rules (credit shared with 1981 Nobel Prize winner Roald Hoffmann), are well known to chemists. As for dissemination of knowledge, Woodward published some two hundred scientific papers and helped train more than four hundred graduate students and postdoctoral scholars (Benfey and Morris 2001, 455–64; Kaufman and Rúveda 2005).

Ciba (Ciba-Geigy from 1970 on) established the Woodward Research Institute to attract Woodward's expertise to Basel, Switzerland, and in the process strengthened ties to Harvard (Eschenmoser 2001). The collaboration worked. Representing Harvard in receiving the Nobel Prize in 1965, Woodward graciously acknowledged his colleagues at the institute. In the early 1970s Woodward completed work between Harvard and the institute on the synthesis of vitamin B_{12}. In the same time frame, he joined Ciba-Geigy's board of directors. Moreover, his industrial interests went beyond large firms. Woodward launched a startup company with other academic scientists and helped commercialize high-performance liquid chromatography in collaboration with the Waters Corporation. Private investments and profits followed the synthetic pathway blazed by Woodward and his collaborators. The engine of growth gathered steam. For example, a standard history of chemistry notes that the "synthesis of the natural cephalosporin was sponsored by Ciba-Geigy, who reaped the benefit by making molecular variants that combined the desirable attributes of both penicillin and cephalosporin" (Brock 2000, 639).

As with the central narrative of this book, a probing, potentially accusatory case study could be assembled, with a spotlight on stock options, missed office hours, potential conflicts of interest, and such. Yet Woodward's success story, both theoretical and practical, has the stronger voice. As a result of commercialization, society as a whole benefited through the enhancement of human health and medicine. Solid results included both organic synthesis of use to industry and a theory of the conservation of orbital symmetry. Finally, the generation and dissemination of knowledge is a matter of record, along with Woodward's own impressive legacy as a teacher (Blout 2001). The minutiae of university-industry interactions settle into a forgotten corner. One is free to assume that Woodward navigated the trials of his industrial collaborations without repercussions. In total, the narrative does not depict worlds colliding. Conflicts of mission yield to missions accomplished.

In looking for the lessons of UCB-N, it is advisable to take into account the Woodward Research Institute and similar episodes. Moreover, concerns popularized in the wake of the agreement should be weighed against the sustained chorus of approval in some quarters for public policy initiatives, such as the Bayh-Dole Act. For example, U.S. Senator Birch Bayh proudly listed a cancer drug (Taxol), a DNA sequencer, and a prostate-specific antigen test as some of the fruits derived from university-industry collaborations (Bayh 2004, 6). One could add Berkeley professor Wendell Stanley's commercialization of the ultracentrifuge for pharmaceutical use and Bruce Ames's namesakes, the Ames Test and Ames Test II for determining carcinogenicity (Kay 1986; Creager 2002; Mowery et al. 2004). It is equally important to remember that many unsuccessful research endeavors are part of the process. With UCB-N, Novartis pursued a research agenda based on prevailing scientific and business assessments. As might be expected in the high-risk enterprise of scientific research for commercial gain, some of the corporation's efforts turned out poorly, but others proved viable. A 2005 article in *Fortune* magazine reported that Novartis led all other firms in the quantity of new drug approvals (Stipp 2005). Given such notable accomplishments, is society endangered if university-industry alliances are shackled in the name of caution? Does the divestiture or recusal of key actors or institutions serve to slow social benefits if individuals with pertinent knowledge are left at the margins?

RHETORIC AND ETHICS IN PRACTICE

Before moving forward to the question of ethics in the Berkeley experiment, an assessment of the norms that underlie the prevailing rhetoric will be useful. Of specific interest is how actors deploy the basic/applied dichotomy as a pivotal rhetorical device. Kitcher (2001, 91) defined the "myth of purity" as "the claim that gesturing at the absence of any practical intent is enough to isolate a branch of inquiry from moral, social, or political critique." Putnam (2002, 2004) puts additional pressure on traditional divisions between objective facts and subjective values. These are not simply findings external to the scientific community, but also the autobiographical lament of scientists in the throes of mainstream research. Microbiologist and Nobel laureate Salvador E. Luria (1912–91) put the same topic forward, recounting the views of some of his fellow researchers. For these "scientific purists," Luria concluded, "the sense of value is often reinforced by the coupling of scientific work with the comfortable expectation that it may actually benefit humanity." Yet science "provides a sort of absolution from responsibility for any controversial matters that may arise in the applications of science. [The priesthood of science] automatically takes science out of the domain of morality by proclaiming it to be intrinsically moral. For the believer in the absolute Good of science anything that goes wrong can only be the fault of the users of science, not of the producers of absolute Knowledge" (Luria 1984, 202–3).

Whether myth or absolution, distance from realpolitik has not dislodged the utility or cultural legitimacy of the basic and applied categories. On the basis of interviews with biologists and physicists in the United States and United Kingdom, a 2001 study found the term "basic" (also represented by the similar labels of blue-sky, curiosity-driven, fundamental, and pure research) to express an ideal rather than a reality for scientists. In practice, scientists find the basic/applied dichotomy "useful when there are specific external pressures to describe research in a certain way" (Calvert and Martin 2001, 19; Calvert 2004). In contrast, while a visiting professor at Berkeley in the early 1960s, British physicist Charles P. Snow warned more forcibly that the "invention of categories" such as those needed for the neutrality-of-science defense was a "moral trap ... one of the easier methods of letting the conscience rust" (Weaver et al. 1961, 133).

The UCB-N controversy offers a fine illustration of the muddle at the basic/applied divide. The closing paragraphs of the *Atlantic Monthly*'s cover story pulled from the past to explore right conduct. How prominent scientists like Nobelist Paul Berg once enjoyed greater autonomy forms the baseline for comparison: "Today, as the line between basic and applied science dissolves, as professors are encouraged to think more and more like entrepreneurs, a question arises: Will the Paul Bergs of the future have the freedom to explore ideas that have no obvious and immediate commercial value? Only, it seems, if universities cling to their traditional ideals and maintain a degree of independence from the marketplace" (Press and Washburn 2000, 54).

The Berkeley-Novartis agreement appears to come up short relative to a historical standard, with a disappearing line between "basic" and "applied" a contributing factor. But the golden past in this comparison is less golden than it looks through the haze of nostalgia. Several million people vaccinated between 1955 and 1963 received an injection of polio vaccine that contained a live monkey virus. Discovered in 1961, the virus, labeled SV_{40}, tested positive as a carcinogen. Intensive scientific research ensued, with scientific findings tightly coupled to the pharmaceutical firms that produced and distributed the vaccine.[1] Unfortunately, research by Berg and his colleagues did not yield a medical remedy. It is understandable that the *applied* element of this research in the formative years of biotechnology quickly became forgettable. The narrative of "basic" science won out, and in this instance serves as a misleading baseline of comparison.[2] In fact, both the SV_{40} episode and UCB-N provide robust examples of the myriad influences on scientific endeavors.

Any assertion that, at the launch of UCB-N, PMB or Berkeley as a whole was an isolated preserve of "basic" research would also create a dubious baseline for comparison. The application of science is an unmistakable part of Berkeley's history. A few highlights from the life sciences should suffice in providing this perspective.

The University of California added a San Francisco-based medical school in 1873, and state support addressed various public health concerns through the State Hygienic Laboratory at Berkeley beginning in 1906 and a Department of Bacteriology and Pathology beginning in 1911 (McClung and Meyer 1974, 261). As before World War I, the interwar period shows a close marriage of science and application. Karl Friedrich Meyer (1884–1974) arrived at Berkeley in 1913. He helped establish research aimed at public health and chaired the Department of Bacteriology from 1924 to 1945. His work encompassed threats to human and animal health such as botulism, brucellosis, leptospirosis, plague, and psittacosis (Bibel 1995, 11–12). Industrial patrons included the California Canners Association and the National Canners Association. These trade organizations helped sponsor a UC laboratory dedicated to botulism and administered by Meyer beginning in 1926 (Cavanaugh 1974; Steele 1974).

Albert Paul Krueger (1902–82) joined UC in 1931 and led the establishment of the Naval Medical Reserve Unit at Berkeley in 1934. The unit explored a range of applied topics related to human health and biological weapons (Cochrane 1947). The U.S. Congress sought to address agricultural surpluses during the Great Depression through the Agricultural Adjustment Act of 1938. As a result, the Agricultural Research Service's Western Regional Laboratory was started near the UCB campus in Albany, California. Even in the early days of biotechnology, when asked to comment on its role in basic research, the USDA emphasized the applied nature of agricultural research: "Within the agricultural system, basic research is generally inseparable from other research in both planning and conduct. This system is highly decentralized. Much research of individual scientists is predominantly problem-oriented and usually developed from a timeframe of five years or less" (National Science Foundation, National Science Board 1978, 4).

After World War II, Wendell Meredith Stanley (1904–71) joined Berkeley as professor of biochemistry and director of the virus laboratory. Stanley informed prospective patrons that his new lab could deliver on cures for cancer, chicken pox, influenza, measles, mumps, and polio, as well as improve protection for agricultural plants and animals (Creager 2002, 154–60). The Biochemistry and Virus Laboratory included the Department of Agricultural Biochemistry and the Department of Bacteriology starting in 1951.

These examples, in addition to agricultural outreach stemming from the land grant mission, fall before the launch of Sputnik in 1957. The cold war and the rapid commercialization of semiconductors and biotechnology continued the historical pattern of science shaped by the promise of application. In 1971, for example, a UC Berkeley discovery of genes linked to nitrogen fixation enhanced speculation that genetic engineering might tackle world hunger. The researchers concluded: "We hope that by applying molecular biology to problems in agriculture, mankind will ultimately benefit."[3]

Nearby Stanford University, a key player in the biotechnology saga, was no less applied in its orientation to the life sciences (Vettel 2004). Established in 1885 through an act of private philanthropy, Stanford moved its medical school from San Francisco to Palo Alto in the late 1950s. The vision for the new facilities was to unite clinical research with new discoveries in the sciences of biochemistry and genetics. The new facilities were designed for a more collaborative environment between medicine and scientific research (Rosenthal 2000). Once again the desire was not to drive a wall between science and application, but just the opposite. Stanford's heavy recruiting for the new medical campus brought talented researchers who proved decisive for biotechnology. Key arrivals in 1958 and 1959 included Charles Yanofsky from Western Reserve University Medical School, Joshua Lederberg from the University of Wisconsin, and Arthur Kornberg and Paul Berg from the Washington University School of Medicine. Kornberg, a physician, had served as chair of the Department of Microbiology at the Washington University School of Medicine beginning in 1953. In sum, in the history of academia in the San Francisco Bay Area, external, applied, and market influences have been an ever-present reality.

Regardless of ties to the applications of science, the allure of "basic" science is easily, quickly, and powerfully reintroduced by scientists in the face of predictions of negative repercussions or the arrival of unexpected consequences tied to new sciences or technologies. For example, the U.S. Congress held hearings in 1977 on genetic engineering; testimony addressed both the promise and the ethics of the emerging field. Paul Berg testified that the "ability to isolate pure genes puts us at the threshold of new forms of medicine, industry, and agriculture" (quoted in U.S. Congress, Senate Committee on Commerce, Science, and Transportation 1978, 36). Nevertheless, he continued, "There are many difficult and contentious scientific, ethical, and moral questions to be examined and at many stages there will be opportunities by all segments of our society to have their say. But preventing or slowing basic genetic research now, seems ill-suited to dealing with that question. We need to distinguish between the acquisition and the application of knowledge" (37).

As questions mounted, scientists deployed the basic/applied dichotomy as needed to table concerns and allow their research agenda to continue unimpeded. Dorothy Nelkin, from the Program on Science, Technology and Society at Cornell University, informed the same Senate committee that scientists had engaged in regular "linguistic manipulations" in the process of making these points (quoted in U.S. Congress, Senate Committee on Commerce, Science, and Transportation 1978, 392).

More than twenty years later, on the verge of UCB-N, the same rhetorical device was used to shield the sequencing of the *Arabidopsis* genome. A plant in the mustard family, *Arabidopsis* became a popular research organism in the 1980s. One reason for its selection is its applicability to oilseed crops such as soybeans. However, the list of potential applications grew to include alternative fuels, chemical (non-petroleum-based) feedstocks, and plants for cleaning up contaminated soil at former nuclear weapon sites (Moffat 1992). Although eager to embrace a wide range of benefits and noting that the proposed research was designed for near-term applications, an important government report artfully dodged larger controversies for the task at hand: "Just as in other aspects of genome science, plant genetics has a number of ethical, social, legal, and economic issues of interest to society. These issues are being addressed on a wide scale, both here in the United States and abroad. Most of these issues relate to the genetic engineering of crop plants, not to genomics research *per se*" (National Science and Technology Council, Committee on Science, Interagency Working Group on Plant Genomics 1998). Here, basic genomics research per se is claimed to partition basic science from applied "ethical, social, legal and economic" issues to defend science and scientists at the same time that the partition is used to declare the ethics and interests immaterial to the justification of science.

Now that we have set the stage, we can look at the ethics of the Berkeley experiment in clearer perspective. Although the agreement was said to be "enmeshed in the highly contentious and emotional political struggle over genetically modified food," the internal reviewers avoided the larger ethical discussion (Price and Goldman 2002, 41). Despite noting that the distinction between "basic" and "applied" research was a "vexing matter," the reviewers, as is customary, relied on the demarcation to arrive at favorable findings, and found "no noticeable movement in PMB's research agenda toward 'applied research,' as was widely anticipated. Rather, there is a marked continuity with respect to the basic subjects of PMB's scientific inquiries" (Price and Goldman 2002, 39).

Simply put, the administrative reviewers placed the actors and activities associated with the agreement within the traditional safe harbor of scientific conduct. As a result, the door on ethical discourse, as it pertains to institutional actions, swung shut with little effort. A similar rhetorical frame can be found among the PMB faculty. Interviews from both the internal and external reviews refer to an interest in funding from the National Science Foundation as proof of the investigators' orientation to "basic" science. Once more, the claim of purity relies on a faulty standard of comparison. For example, one NSF proposal nearly identical in subject matter and timing to the agreement stated as its aim a mixture of applied and basic concerns related to "insights into basic plant biological processes and the ability to genetically manipulate plants for agronomic improvement."[4]

While heroic narratives of science paint many a golden past, closer examinations routinely reveal the unsteady hand of patronage and a variety of other external influences. In practice, ties of patronage and socioeconomic promise appear an unavoidable necessity. The working assumption that universities should keep to "basic" research and industry to "applied" efforts is simply rhetorical. As Yost (2003, 35) concluded, conflicts cannot realistically be removed from campus, for they are "inextricably bound to complex activities of research universities and their extramural sources of financial support." A proud and historical convention among academic scientists is their allegiance "to the advancement of knowledge for its own sake, and not for the sake of money or fame, or of professional positions or advancement ... [undertaken] not with any reference to the probable utility of the results." Yet the limitations to this maxim seem obvious, forcing an equally historical contradiction to allow for "making other lives more pleasant, of bringing light into the dark places, of helping humanity" (Billings 1886, 543). To this desire one must add the public's expectations, raised in part by scientists themselves. Glass (1965, 101) pointed out that within the public university, the scientist has "eagerly grasped for social support, and ... thereby acquired social responsibilities." Bringing benefits to humanity often calls for a carefully constructed bridge to the financial resources of the commercial sector with the associated trials and responsibilities.

The constraints of practicality can also be seen in the context of intellectual property and human mobility. The debate over intellectual property rights under UCB-N may leave the impression that knowledge flows from scientific literature or conversely from patents and legal agreements. In a truer sense, the transfer of knowledge is about people, who can and do move from place to place (Shane 2002; Cummings and Teng 2003). Consider a few careers relevant to the scientific path of UCB-N. Robert T. Fraley and Robert B. Horsch took their 1979 doctorates from the University of California, Riverside, into fruitful careers with Monsanto. Mobility can also occur at mid-career. Mary-Dell Chilton received her doctorate from the University of Illinois in 1967 and joined the Department of Microbiology and Immunology at the University of Washington in Seattle. In 1979 she moved to the Department of Biology at Washington University in St. Louis, Missouri. Her formal industrial career began in 1983 when she joined the Ciba-Geigy Corporation (a Novartis and Syngenta precursor).

Other scientists craft careers with an entrepreneurial flair. Barry L. Marrs, who earned his doctorate in 1968 from Western Reserve University, was a postdoc under Yanofsky at Stanford (Marrs 2002). He became a professor of biochemistry at St. Louis University before launching an industrial career in the mid-1980s. After nine years with DuPont, he founded Recombinant Biocatalysis, Inc., in 1994, which later became Diversa Corporation. Central to the UCB-N case, Steven P. Briggs followed a 1982 doctorate from Michigan State University with assignments with Pioneer Hi-Bred International, Cold Spring Harbor Laboratory, TMRI, and Diversa (Vogel 2002). In 2004 he became a professor of cell and development biology at the University of California, San Diego. Given the perhaps rising frequency of careers such as Briggs's, those proposing tighter constraints on university-industrial relationships—usually higher and more impenetrable walls—should take into account the degrees of freedom associated with knowledge-laden mobility within and across historical and contemporary university-industry relations. A strict university/industry demarcation, like a basic/applied one, may prove a shaky foundation.

In an environment where academic and industrial interests blur, the promotion of university-held and industry-licensed intellectual property rights aims for a concise demarcation where no crisp division exists, thereby generating the widespread and lingering uncertainties expressed in much of the resistance to UCB-N. While the academic-industrial boundary is repeatedly crossed, institutional approaches to IPR and the agreements to manage them are crafted as if establishing a bright line of ownership and control. A few historical examples highlight the range of options and ensuing questions.

In the late nineteenth century a collaboration developed between the Case School of Applied Science (one precursor to Case Western Reserve University) and a new startup company. Case's president, Cady Staley, invested in the firm with the insurance that the company would capture Herbert H. Dow's intellectual property with the aid of Case professor Albert W. Smith. Smith and Staley joined the company's board of directors at its founding in 1897. In time, one observer noted that the Case School became "virtually an extension of The Dow Chemical Company" (Whitehead 1968, 72). Does the long-term success of this corporation justify the intrusion on academia? If the relationship contributed to the undeniably good reputation of Case Western, ought it to be viewed as an intrusion?

In the late 1920s Roger Adams chaired the Chemistry Department at the University of Illinois at the same time that he collected an annual salary from DuPont. Adams's agreement required that he consult exclusively with DuPont and sign over his IPR as part of the bargain (Hounshell and Smith 1988, 228). Is the professor liberated by his lack of interest in patents, or is the public shortchanged? In the late 1940s computer pioneers John W. Mauchly and Presper Eckert's University of Pennsylvania employer clamped down on IPR and demanded that university interests trump commercial pursuits. As a result, the researchers left Penn to launch Eckert-Mauchly Computer Company, a forerunner of Unisys (Pugh 1984, 15). Was the university better off with their departure to industry, or were the ranks of academia diluted? Are these questions largely immaterial if the outcomes are deemed successful for the university, for the economy, or for society? The potential for success, as well as the definitions of success, do bring additional uncertainty and anxiety to the trials of university-industry alliances. Among the tests are ethical choices made or deferred.

CONDITIONS FOR THE EMERGENCE OF RESEARCH CONTROVERSIES

We turn now to expanding and generalizing where appropriate on the recommendations provided to Berkeley through the external review. Given the experience of UCB-N, how should universities handle future relationships with industry? How can the core principles of academia be protected under the weight of external pressures? The Berkeley experiment occurred in a tempest of external anxiety over genetically modified organisms. On a broader scale, consumer, corporate, organizational, regulatory, judicial, executive, and legislative decision makers look to university investigators for impartiality on critical topics. Decisions on many concerns facing modern societies include grappling with technical questions on the safety of food, water, air, and medicines. Should this product be allowed on the market? Should that drug be banned? Are illnesses traceable to exposure to this chemical? Is anyone liable for that danger to public safety? Is this new technology a threat to the environment?

While the UCB-N case provides a constructive springboard for investigating universities in general, more conditional generalizations can offer further guidance for practitioners confronted with tangled missions (George and Bennett 2005). The existence or emergence of similar situations may be broadly related to four interconnected conditions:

1. Contested science or knowledge claims illuminated by public controversy or regulatory uncertainty.

Many of the daily events at universities fall outside this first condition. Academics can reasonably expect little controversy over teaching calculus or conducting research on sixteenth-century French literature. Different expectations prevail for a subset of controversial topics, such as global warming or stem cell research. This first condition would appear easy to identify given the open nature of public controversies. Activists do find their voice in popular, scientific, and other professional media. A controversy's resonance on campus is also relatively transparent to an attentive observer. The public controversy surrounding UCB-N was not a sudden or unexpected event. Rather, university administrators were very much aware of the gathering storm and had umbrellas and galoshes ready. At the same time, the historical struggles over agricultural science and policy in the region armed opponents with ready thunderbolts. In other cases, it may not be apparent to the public that the various missions of the institution are at odds. Other scenarios could include regulatory concerns that surface within a professional community well before it reaches public notice. Greater scrutiny would be needed to identify such events in a timely fashion.

2. Potential findings or expert advice stemming from university endeavors with near-term relevance and significant consequences beyond academia.

This condition speaks to the relevance and immediate appetite for academic outputs. Research on a looming epidemic can be distinguished from the latest thinking on the Big Bang given the clear anticipation in the former case of corporate, governmental, or public response. As with the first condition, external demand for comment and expertise constitutes a distinct flag. Easily recognizable indicators seem likely, among them external activities relating to patenting, litigating, assembling venture capital, forging regulatory decisions, and debating public policy. The UCB-N examples include the multiple missions of promoting genetically modified organisms, the appropriate regulatory regime for GMOs, and alternative approaches for the future of agriculture. Scientific counsel from research under way and within the agreement moved quickly and directly to decision makers external to the university.

3. Institutional or administrative involvement coupled with a proactive stance held by key decision makers with institutional authority.

The first two conditions may be met, yet university administrators may be disengaged, having no stance or direct interest in the controversy. For example, a public debate may rage over the earthly danger from asteroids, with the U.S. Congress holding hearings and seeking scientific counsel. Nevertheless, university administrators might remain uninterested in the topic. Conversely, with institutional management of intellectual property, this condition may be broadly present, since knowledge creation may be converted to licensing revenue for the institution in an unpredictable fashion. A bright-line test is feasible here as well, since public statements on behalf of the university typically accompany its institutional missions. For example, the University of California openly promotes the economic potential of biotechnology.

4. The use of public funds, separately or intermingled with other resources, for knowledge creation, interpretation, or dissemination.

Given known controversy and the institution itself being a stakeholder in immediately relevant outcomes, the fourth condition to be met is the existence of public funds. Is the institution accountable to the public through the public purse? With standard academic accounting, this is to be determined. In some isolated instances, even at public universities, the private sector does pay as it goes without the consumption of public goods. For the majority of other activities, the public has a financial stake in academic outputs.

Since these four conditions were met in the UCB-N case, it takes on additional significance. Such conditions require that academic leaders should be ready with flexible approaches to handling ethical issues that look well beyond the lone investigator (e.g., Edsall 1975; Martin and Reynolds 2002; Lefkowitz 2003). Methods should be transparent to stakeholders and thoughtful enough to earn the respect of the most creative academic minds. Implementing effective approaches will demand attention to internal boundaries and conventions in addition to the external forces at play.

INTERNAL BOUNDARIES AND CONVENTIONS

Any analysis that treats UIR as a connection between two black boxes sidesteps a sea of complexity. In the UCB-N controversy the lid came off the black box labeled *university* to reveal multiple voices talking past each other. Even more intensively than in Clark Kerr's multiversity, the controversy over the Berkeley-Novartis agreement makes apparent that the aggregations of a university's interests, roles, or norms as a monolithic constant—one easily contrasted with that of industry—is overly simplistic intellectually and utterly contradictory in practice. Rather, one finds running debates between widely and differentially situated actors seeking to forge a working consensus or an isolated autonomy.

In the UCB-N case, day-to-day outcomes in the public sphere emerged in part from individuals and institutions with manifold administrative duties, including California state legislators, the UC board of trustees, the UC president, the chancellor, the CNR dean, the PMB chair, and the principal investigators. To fully grasp the internal density on campuses such as UC Berkeley, add to these decision makers the Academic Senate, more than a dozen different committees, representatives from other academic departments, other administrative units (e.g., intellectual property, sponsored research, legal affairs, public relations), as well as individual faculty, students, and staff.

While the UCB-N interviews captured local reflections on specific personalities and positions, a broader view sees participants as acting out their distinct roles within the institution. As Hatakenada (2004) implies, the internal boundary between academic researchers and university administrators may be just as dynamic as the university-industry boundary. In practice, the career rewards of research investigators are closely associated with specific disciplines and the responsibilities of research, teaching, and service. The roles of administrators, by contrast, tend to be more closely linked to the financial, regulatory, legal, and political accountability of the institution. Administrators typically have authority over bureaucratic techniques and tangible resources used to serve institutional objectives. Investigators who compete successfully for extramural support must engage administrators on many fronts to turn their awards into expendable funds. There are also key differences in the professional incentives for faculty and university administrators.

In this way, not only are there institutionally embedded political differences when it comes to the normative and ethical conventions upon which different faculty draw in public controversies about the mission and methods of the university, but equally embedded and different conventions exist between faculty and administrators. In the latter relationship, the populist-progressive split does not illuminate the dynamic as it does the former. On the one hand, rather than a rigid hierarchal arrangement, university administrators take on something like a strong-client relationship with investigators. A funded project does not take place without a contractual event between the sponsor and the institution (e.g., a sponsored research agreement). The tasks at hand for administrators often call for responsiveness to leading patrons and attention to compliance, cohesion, and accountability. Administrators may seek productive ends by encouraging team approaches to knowledge production and innovation, hoping to provide motivation for some investigators without alienating others. The predictable result is a range of disputes and alliances across internal boundaries. Administrators thus take on institutional burdens related to compliance and consistency in the face of a complex assortment of policies, guidelines, regulations, and legal statutes.

On the other hand, university administrators should not be depicted as working against autonomy, creativity, or diversity, but they do shoulder responsibilities that may constrain as well as advance these principles. This posture is not without prior justification (Olswang and Lee 1984). Universities have their own records of actions well outside accepted norms. Strewn across the terrain of accumulated administrative policies is a (usually) hidden legacy of the occasional individual engaged in capricious, deceitful, or outright reprehensible behavior. Nevertheless, the key is the different norms, ethics, and conventions rooted in various locations across the university and the ways these differences generate complex, if not incommensurable, individual and institutional strategies.

In general, behavior in universities may be seen as a loosely interrelated collection of conventions. The UCB-N case highlights several identifiable entities in the thick of the action, including PMB, ESPM, SPO, OTL, OCG, and UCOP. Each entity acts within its own unique set of conventions. As academic departments, and despite their differences, ESPM and PMB share much in common, as both are research-based units. At the same time, they differ on specific research agendas and perceived pathways to public goods. As depicted in their businesslike annual reports, the administrative units SPO and OTL both tally financial indicators as signs of success. However, goals for SPO may differ from those of OTL when intellectual property provisions in research contracts appear overly restrictive. Moreover, efforts to expand the financial pie may run into risk-averse legal reviews from the OCG. Finally, UCOP may be focused more strictly on public perceptions, long-term financial stability, and the realities of external politics. In sum, internal boundaries between investigators and administrators, as well as between various administrative units, have important distinctions relative to conventions and conduct.

Investigators—Individual Integrity

For an academic researcher, arriving at proper conduct is not a simple exercise in knowing what one's institution expects or how it interprets state and federal requirements. The investigator faces a range of situations that fall both inside and outside the university's reach. Beyond the wide range of campus activities are other domains that are equally important to the role of the modern academic researcher. What are the ground rules of presenting a seminar for the private sector, of consulting for industry, of giving testimony before the state legislature or the U.S. Congress, of reviewing proposals for government funding, of serving on a scientific advisory panel for a government agency, of acting as a member or leader of a professional society, of reviewing the work of a peer for a scientific journal, of giving expert testimony in a court of law, of speaking on the record to a journalist, of exercising the right of any citizen to address the public at large? The answers to these questions differ and may also diverge from acceptable behavior within the confines of the scientific method or under the legally binding language of sponsored research agreements. Answers may also have valid differences within peer relationships and faculty-student relationships, or they may be distinct in the classroom, laboratory, boardroom, and public discourse.

The ethical bounds for these interrelated tasks, even when codified, are not unified, easily summarized, or intuitively obvious. Within major research universities one is likely to find complicated overlapping policies and guidelines. For example, the University of California released a formal guidance on consulting in March 2003 (University of California, Office of the President 2003). University leaders did not have a new policy to report. The document aimed only to simplify the boundaries of acceptable conduct for faculty engaged in the common practice of consulting. Nevertheless, the guidance covered seventeen pages and referred the reader to eleven university documents, three government agencies, three state statutes, and a federal law. Given that consulting is one small piece of a scientist's domain, the overall complexity of right conduct is disconcerting. Learning from UCB-N in order to add yet another policy or guidance to this mix does not appear constructive, unless a more global effort were able to enhance overall comprehension. The University of California is by no means alone in this labyrinth. Truly simplified written standards of conduct are widely needed. New guidelines should provide clarity across multiple roles, institutional settings, and missions. New efforts should seek to supersede standing sets of oddly overlapping policies and practices.

Administrators and investigators should also appreciate that conflicts of interest and ethical choices are not fully contained in the four corners of legal contracts. For the institution, remunerative avenues for investigators that could be sources of conflict, such as stock ownership or options, royalties, gifts, honoraria, travel and entertainment, lucrative consulting engagements, and other fees are relatively easy to detect and manage. However, the procedural convenience provided by a flow of documents should not be mistaken for attentiveness to a dynamic human process. The current mainstay of financial disclosures offers only a flag to decision makers who may or may not be motivated to respond, and even when disclosures go forward they may not achieve their intent.[5]

In addition, a wide range of nonfinancial incentives may stem from traditional academic interests, including achieving tenure, collecting citations, placement on high-profile committees, obtaining administrative powers, or academic advancement. Such conflicts may prove more difficult to detect. Even advocacy from one's discipline or research agenda, the unselfish promotion of one's institutional home, or loyalty to one's students may tip the balance without attention to personal finances. Just the drive to be first to publish in the pursuit of good science or for a chance at fame may be enough to motivate questionable behaviors. For example, Korn (2000, 2234) finds conflicts in biomedical research to be "ubiquitous and inevitable," stretching beyond financial conflicts to encompass the "desire for faculty advancement, to compete successfully and repetitively for sponsored research funding, to receive accolades from professional peers and win prestigious research prizes, and to alleviate pain and suffering."

The result can be a "strong bias toward positive results." The traditional litany of scientific misconduct includes falsifying data, selective reporting, breaching confidences, plagiarism, and circumventing safety regulations. Clearly written policies, along with active dialogue to keep mental models of right conduct fresh and relevant, are necessary to inform daily behavior. Additional consideration of nonfinancial COI stemming from allegiance to a specific discipline or research agenda is also needed.

Administrators—Integrity and Accountability

With expanding management of IPR and economic development agendas, institutional conflicts of interest need heightened scrutiny. As discussed above, the UCB-N case study suggests that the boundaries of current COI policy and codes of conduct are too narrow in several respects. For an institutional process to find legitimacy, COI polices, long targeted toward faculty investigators, should extend to the institution as well. One direct approach is to avoid industry agreements that involve complete academic units or large groups of researchers. UCB-N's coverage by one firm of nearly all the faculty members in one department was noticeably outside the mainstream of research contracts with industry. While an intriguing experiment, there appears little rationale for repeating the approach. Standard agreements or templates can serve to streamline negotiations with industry without the complexity of unit-wide decision making. With general terms agreed upon, individual research projects may be defined around specific researchers and the specific scope, scale, duration, and matching funds involved. Yet, while easier to administer, such an approach is only a partial solution.

At the institutional level, impartiality and advocacy make an awkward couple. Even in the realm of human medicine, where management of conflicts is well advanced, institutional conflicts fester beneath the surface. Johns, Barnes, and Florencio (2003, 742) found that research integrity may be "compromised by the actual or expected pecuniary corporate interests of the institution or its departments" by "noninvestigator institutional decision makers."[6] University leaders may be failing to fully grasp a growing need for institutional accountability. A university can harbor individuals with different ethical and professional interpretations and can proudly do so in the name of diversity and academic freedom. However, university leaders cannot cloak the university as an institution in the academic freedom designed solely for individuals without creating a form of institutional autonomy short on accountability. Killoren (1989, 1) offers a useful statement of principle in this context: "No matter how worthy one might judge the enrichment of one's own institution or how many gains one might project for society as a result of the enrichment, a university cannot betray its basic principles to gain financial security."

Without ignoring them, of course, universities must strive to be independent of their own financial interests when it comes to enabling academic autonomy, creativity, and diversity. A way forward for academia should include universities as institutions that stand firmly for open forums of forthright dialogue on the myriad means to the ends of autonomy, creativity, and diversity. A clear lesson of the Berkeley-Novartis agreement and much of the literature on the state of the contemporary university is that these cannot sustainably embrace singular means of attaining the goals of core principles, particularly when those principles are unrecognized, unaddressed, and incommensurable and are pursued by actors across the university following institutionally divergent patterns of self-interest.

In other contexts, considerable experience exists with standing or ad hoc committees to labor over these kinds of dynamic institutional issues. From the administrative side, university leaders could establish new working panels with advisory or decision-making responsibilities limited to the available discretion of noninvestigator administrators.[7] In practice, methods would need to be customized to specific organizational environments. For simplicity, a single panel arrangement is discussed here. The panel could focus on how the institution can best "keep the public trust and maintain institutional independence and integrity" (University of California, Office of the President 1989).

Given the socioeconomic importance of universities, the first order of business in recommending change is to do no harm. The strength and promise of a committee approach typically rests on a well-defined and limited scope. In this instance, the panel's role is not one of finding virtue or prioritizing institutional missions. It should not serve as line judge on a fact/value, basic/applied, objective/subjective, Humboldtian/Kantian, Baconian/Mertonian, populist/progressive, risk/benefit, or any other similar spectrum. The panel should not look to weigh the relative merits of disciplines, departments, or research agendas. With the exception of those enabled by their employer to act as agents of the institution, the panel should refrain from more audits aimed at investigators that place additional pressure on core principles. Also out of scope would be cataloguing stakeholder representatives, as well as encouraging or placating any subset of diverse publics. In the same vein, orchestrating public involvement or interdisciplinary interactions to resolve conflicts should be placed out of bounds. The panel should not opine on whether new science or technology will spur or disrupt commerce, enhance or jeopardize national security, strengthen or weaken public safety. The panel need not anticipate societal implications or attempt to head off unintended consequences. All of these topics are larger than the institution, and these matters may be left to lay and professional judgment in larger public forums, where university actors can freely participate.

Finally, the panel's domain should not be one for building better walls. Like the unicorns of mythology, scientists working without external influences have been revered for centuries, but never actually sighted. Building higher fences to better shield an ideal habitat for unicorns, sadly, will not result in the elusive creatures' returning to prance in the early morning dew. The same disappointment will await those who fortify firewalls so that pure scientists may return to an open commons of unsoiled objectivity. Rather, as Williams-Jones (2005, 250) concludes, practitioners are left with "the task of determining what *forms* of commercial influence are appropriate for enabling universities to develop knowledge that is socially, politically and economically relevant."

Having suggested what it may wish to avoid, what should such a panel hope to take on as its charge? First, it should aid noninvestigator decision makers in avoiding any conscious or unconscious actions that suppress informed debate and public communications. Proper questions might include:

- Has institutional advocacy crossed the line to appear as an institutional endorsement at the expense of dissenting opinion?
- Is the institution casting an illusion of objectivity or unity at the expense of diversity?
- Is the institution openly representing and supporting its diversity of expertise even when some investigators hold views deemed out of step with institutionally led objectives?
- Are there inequalities in the enforcement of policy or selective access to public relations, legal, fund-raising, or other administrative competencies?
- How best can the institution balance diverse internal interests given its limited autonomy over resources and policies?

Drawing from familiar U.S. government codes, for anyone in the academic community, a simple test query might read:

Given the situation at hand and conceivable consequences, would a reasonable and well-informed person question my or my institution's impartiality, should this relationship be formalized?

Any member of the academic community who answered in the affirmative could trigger a response from the panel.

The panel's institutional focus should not discourage explorations into the university's larger sphere of influence. The walls that bound the university are characterized by their permeability. Given the mobility of actors across boundaries and the financial, political, and in-kind support to various boundary

TABLE 12.1   Key Questions Regarding Potential Institutional Conflicts of Interest

Are the following four basic conditions present?

1. Contested science or knowledge claims illuminated by public controversy or regulatory uncertainty.
2. Potential findings or expert advice stemming from university endeavors with near-term relevance and significant consequences beyond academia.
3. Institutional or administrative involvement coupled with a proactive stance held by key decision makers with institutional authority.
4. The use of public funds, separately or intermingled with other resources, for knowledge creation, interpretation, or dissemination.

If so, is there evidence of institutional:

1. actions that suppress informed debate and public communications?
2. advocacy giving the appearance of an institutional endorsement at the expense of dissenting opinion?
3. actions creating an illusion of objectivity or unity at the expense of diversity?
4. actions constraining those holding views deemed out of step with institutional objectives?
5. inequalities in the enforcement of policy or selective access to public relations, legal, fund-raising, or other administrative competencies?
6. awareness of reasonable and well-informed persons questioning their own or the institution's impartiality?
7. actions improperly influencing commercial, governmental, and other organizations within the institution's sphere of influence?
8. actions that might dilute the implementation or enforcement of desirable regulations considered inconsistent with institutional or private-sector interests?

organizations, the panel should be prepared to test the validity of assumed firewalls where they are designed and relied upon to limit institutional accountability.

The panel should review the full extent to which academic actors, especially administrators, may influence commercial and governmental affairs as employees, consultants, or avocationally (see Table 12.1). Various nongovernmental organizations operate within the sphere of influence of major universities. Trade, professional, economic development, and special interest organizations are often important conduits of scientific findings under public, regulatory, and legal review.[8] Many of these entities can and do have an impact on the private sector above and beyond direct UIR. University actors can exploit these organizations to promote institutional missions by indirect means. Boundary organizations may focus on straightforward advocacy or distinct conventions adhered to by academic administrators, academic researchers, and commercial actors (Jasanoff 1990; Rouse 1992; Guston et al. 2000; Cash 2001). Therefore, the panel should look at institutional and individual codes of conduct that address relations with organizations linked to the university and its faculty, students, and staff. The panel could also ensure that organizations associated with the university or supported by institutional resources are disclosed and their significance noted. Involving external representatives on the panel could enhance this perspective and include individuals loyal to the institution's reputation but not beholden to or dependent on it.

Early on in the UCB-N case, La Porte (2000, 76, 80) recorded the concern that the scope of research at a public university should aim "not only to develop new agricultural products and methods, but to provide insights into the consequences of their widespread deployment." The danger to the public is that university-industry bonds could dilute the implementation or enforcement of desirable regulations considered inconsistent with private-sector interests. Financial return from IPR stems from new or improved products and services. However, these new offerings must clear regulatory and legal hurdles. Since university technology transfer offices release IPR without warrant or liability, the burden of responsibility for the consequences of new technology falls on local, state, federal, and international regulatory agencies, as well as on the legal system. For innovations related to agricultural biotechnology, licensees of university-owned IPR conduct business in a highly regulated environment, including procedures under the control of the USDA, Food and Drug Administration, Environmental Protection Agency, Federal Trade Commission, and the antitrust authority of the U.S. Department of Justice. These government agencies have been anything but inactive. While government regulators are the final protection against unexpected consequences of new technology, these actors frequently rely on the expert opinion of university researchers. Successful commercialization may enhance the institutions' ability to secure additional industry-sponsored research and may deliver royalty returns. In contrast, scientific findings that lead regulators to bar commercial entry would produce no downstream financial advantage. Thus universities are open to at least the perception of a conflict of interest. For example, facilitating the flow of public resources toward agricultural biotechnology research and away from research aimed at assessing the potential human or environmental hazards of biotechnology could be perceived as a conflict.

An illustration related to the UCB-N case concerns the errant release of corn not approved for human consumption. In 2000 consumers learned that genetically engineered corn known as StarLink had entered the food chain (Durham 2005; Sheldon 2004). The corn contained a protein derived from *Bacillus thuringiensis* (designated Cry9C) that was improperly found in some three hundred products for human consumption. The incident proved costly for the biotechnology industry and spurred regulators into a fast-paced response in the months and years to follow. Uncertainty revolved around testing GMOs for allergenicity (Hileman 2002; Bucchini and Goldman 2002;

Selgrade et al. 2003; Buchanan 2001; Kimber and Dearman 2001). In February 2001 UC investigators filed a patent application dealing with the regulatory uncertainty "raised by critics and other observers of the debate." The patent application explained that "food safety issues raised by the application of gene technology to some foods have yet to be satisfactorily addressed" (U.S. Patent Application No. 09/776,454, 2). Then, between 2001 and 2004 but not publicly known until March 2005, one of Syngenta's transgenic corn varieties not approved for human consumption was accidentally released into the food supply (Macilwain 2005; Schmidt 2005). The debacle delivered another blow to the biotechnology industry, regulatory confidence (especially in Europe), and U.S. trade relations. Scientific advice to government regulators directly affecting Syngenta's errant release involved scientists under UCB-N.

UC as an institution entered into UCB-N with an explicit statement of applied intent. PMB scientists engaged in science with the anticipation of applications transferring into commerce. The institution housed research and produced findings of interest to industry and regulators of industry. Activities cannot be explained away as disinterested science. How can an institution imply science above the fray of public controversies as it gathers intellectual property aimed at addressing the very same controversies? If investigators or institutional interests in commercialization can influence related regulatory or legal restrictions, a conflict of interest, or at least the perception of one, emerges. When it comes to regulatory concerns, individual investigators are already immersed in a highly monitored environment (e.g., the Federal Advisory Act). Yet the same cannot be said for the institution.

It should be made clear that this case study of UCB-N found no misconduct, and even under heightened scrutiny the University of California may well emerge without a blemish. Moreover, close interrelationships are not uncommon in regulatory affairs; some valuable scientific results may not be achievable without close university-industry collaboration. This is more likely to occur when public funds are lacking. For example, with total biomedical and life science funding at $18.2 billion (FY 2001), a report from the Pew Initiative on Food and Biotechnology estimated total federal funding for research specific to food allergies at between $4.2 and $7 million (Bucchini and Goldman 2002, 16).[9] All the same, the UCB-N agreement as an experiment in university-industry agreements provides evidence that improved institutional COI guidelines are needed to better safeguard regulatory procedures and public confidence. When circumstances of this type dictate, a review panel could assess institutional obligations and commitments to reliable production and communication of regulatory science and related scholarship.

An earlier European study found the same tensions between the university mission of "contributing to national economic performance by increasing links with industry" and the mission "to improve risk assessment and regulation in the public interest" (Tait and Chataway 2000, 22). This second mission, as described above, includes protecting the public from the Pandora's box of science and technology. Technology transfer offices at major research universities acknowledge and embrace the importance of knowledge production and dissemination through timely publications (Nelsen 2004). But how the institution shoulders the responsibility of Pandora's box relative to a university's technology-transfer function is in need of illumination. Given the role of universities in society, the panel, like the institution, should not shy away from public controversy. Conflicts over knowledge claims often arise where the stakes are high and academic goods essential. The integrity of regulatory counsel should not be disadvantaged, turned into a have-not compared to the legislative intent of the Bayh-Dole Act and like governmental directives. The important point is that investigators be free to speak out and publish in order to provide the public and regulators with the latest findings on topics of interest. The panel could review the university's financial support of regulatory science and how the institution encourages an environment for open reporting of findings and analysis. The panel should be especially diligent with perceived institutional conflicts of interest on campus that could undermine confidence in a regulatory system associated with emerging science and innovation. Additional notice to the health of regulatory science would be both a public service and a check against institutional duplicity.

Planning is also needed to avoid timelines that could threaten to overtake ethical considerations. The panel's actions would need to be contingent on specific and identifiable situations. However, to be effective the panel would need to establish anticipatory and flexible approaches to take action on timely matters. Its activities would be enhanced by review early in the process of developing research agendas. For example, the overall policy agenda for biotechnology took shape between the late 1970s and early 1990s. By the late 1990s, related commercial activity, with its university and governmental components, had gathered considerable momentum. The UCB-N debate came too late for some participants to exercise meaningful degrees of freedom. Events that tower in importance over UCB-N include the Asilomar meeting on recombinant DNA (1975), the *Diamond v. Chakrabarty* Supreme Court decision (1980), the release of the U.S. government report on a regulatory framework for biotechnology (1986), the release of U.S government policy on foods derived from biotechnology (1992), and the recommendations of the National Science and Technology Council's Interagency Working Group on Plant Genomes (1998).

While the outcomes of UCB-N do not appear pivotal relative to research agendas, earlier arrangements that *were* pivotal included input from UC scientists
and administrators. Therefore, the panel may wish to engage in timely reviews
of institutional commitments to alternatives to the dominant paths of inquiry
in their formative years, especially when faced with public controversy or regulatory uncertainty.

One feature apparent in the UCB-N experiment was the continuation of
a long-term pattern whereby certain participants successfully circumvent public and scientific concerns surrounding a specific research agenda. The panel
could work to identify and prevent the masking of intended applications of
knowledge or potential negative consequences of commercialization with the
privileges implied by academic freedom. There are indications that the over-
statement of academic freedom, mixed with interests in commercialization,
is simultaneously eroding the public trust in science and the genuine merits
of scholarly autonomy. University scientists and administrators have found
occasion to use the ideals of "basic" science and academic freedom to create
distance from accountability relative to any ethical, moral, or economic issues
pertinent to their research agendas.

Academic freedom should encourage the expression of alternative views
rather than shelter conduct from scrutiny. The privileges of academic freedom
do carry the responsibility of professional ethics and candor. In this context,
university actors should confront the rhetorical construction of pure knowledge distinct from external influences. Is it truly meaningful to assert that
resource allocations or tenure decisions are made solely on the merits of science or scholarship? Or does this ideal now serve as the central pillar in a dysfunctional debate? Can the academic community refine a working ideal for
scientific research that is better grounded in the current practices demanded
in applications and by patrons? A revised approach could include administrative procedures and mechanisms for faculty, students, and staff seeking
assistance with professional tensions or multidisciplinary disputes apparently
at odds with one or more institutional missions. Thoughtful assistance would
acknowledge the complexity of outside influences rather than assuming that
these influences may be eliminated with heightened vigilance in the name of
academic freedom or pure science.

The panel envisioned here could be advisory, offering an appraisal to other
decision makers, or it could be empowered in its own right. Its powers could
include the reallocation of resources or future earmarks on funding streams
within the institution's discretion. If licensing revenue leads to self-reinforcing
research agendas based on the potential for commercialization, tapping the
institutional share of these incomes might be warranted. The panel could also

be empowered to require or approve proposals from administrators to avert institutional conflicts.

Finally, proper attention to ethics entails continuous education. On campus, investigators and administrators should have a better grounding in the specifics of intellectual property management and policy in the context of knowledge creation and societal benefits. Universities might profitably assess how they strive to educate the public on the specific nature of intellectual property, technology transfer, and institutional accountability. Much has been made of the public's general lack of scientific knowledge. In the related context of the failings of universities to teach evolutionary science coherently across the natural scientific curriculum and its inability to argue against creationism and intelligent design, Pennock (2003) has convincingly argued that the hiring practices and reward structures associated with privileging research and UIR undermine the kinds of effective multivalent and interdisciplinary teaching necessary for investigators, administrators, regulators, teachers, and the public to understand technoscience—controversial and otherwise. If science is inadequately taught—even at universities—then it is incumbent on researchers appealing to the conventions of basic science, and on administrators invoking the idea of universities as engines of growth, to both better understand the unintended consequences of their own practices and express the merits of their agendas across the university and to the public.

It is also true that there is considerable misunderstanding regarding patenting and university licensing. This lack of understanding may lead to the belief that patented technology developed at and licensed by a university carries a de facto endorsement by the university. Quite the contrary, universities license innovations without warrant or liability. They do not guarantee economic value, ethical use, social benefit, or manageable risks relative to public safety or the environment. The absence or extent of institutional liabilities should be more actively communicated. Universities should also address concerns that industry sponsors of university research may seek only an implied "stamp of approval" for their enterprise rather than demonstrate a genuine interest in the promotion of science and innovation. In such circumstances it is the goodwill traditionally identified with academic creditability that is improperly sought for private gain.

Whether via a new panel as illustrated here or through heightened awareness followed by leveraging available means, the champions of academia must redouble efforts to ensure the highest possible standards of independence, integrity, and candor. Universities will best serve both the public and their own core principles by bringing forward goods in good faith, without censure or bias.

TOWARD A NATIONAL DIALOGUE ON THE
FUTURE OF UNIVERSITIES

The UC Berkeley–Novartis agreement is symptomatic of the problems facing our universities today. While there is widespread agreement on the importance of autonomy, creativity, and diversity, there is by no means a consensus as to what those terms mean. In a very real sense, we are heading into the future facing backward—looking to a past that has become in part mythical and that in any case can no longer be resurrected. While countless researchers subject their written work to scrupulous peer review in countless journals, no one seems to be submitting our common work to anything resembling peer review.

Moreover, the division of knowledge established in the eighteenth and nineteenth centuries—with its separation of knowledge of the natural world from knowledge of the human world, of basic from applied research, of Kantian from Humboldtian traditions, of science from engineering, of art from science, science from technology, teaching from research, of classroom teaching from extension activities, and of public from private knowledge—is crumbling before our eyes. Worse, it is crumbling despite the intensified and endless repetition of "rituals of verification" (Power 1997) that (we are told) ensure what is often vacuously referred to as the pursuit of excellence.

What is needed is a multilevel and ongoing dialogue on the future of universities. Such a dialogue needs to include faculty, students, staff, administrators, and perhaps alumni at every major university. But it cannot be limited to them. It must also include faculty organizations like the American Association of University Professors and the American Federation of Teachers. It must include the academic professional and disciplinary societies. It must include the many university associations, such as the Association of American Universities and the National Association of State Universities and Land Grant Colleges. It must include the federal funding agencies, such as the NSF and NIH. It must include the National Academies. It must include the foundations that support education.

The questions to be addressed by such a dialogue would be fairly straightforward: What kinds of universities do we want in ten, twenty, or fifty years? Should they exist at all? If they should exist, what roles should they play in future societies? How shall we plan on getting there? What roles will teaching, research, outreach, and service play in these universities of the future? How shall we fund these institutions? What models exist to help us along the way? What perils must be avoided if we are to succeed?

The intent of such an endeavor would not be to create some uniform model toward which all universities should strive. That is both an unlikely outcome of such a dialogue and perhaps an unworthy goal. We are far too diverse in our interests, outlook, ethnic and cultural backgrounds to have a one-size-fits-all approach. But we can look to the development of multiple models designed to build different social worlds, each of them in touch with the others.

We live in a time of enormous social, political, economic, and technical change. Universities are arguably the one place in our fast-paced society where reflection on purposes is still possible. Yet, ironically, we have become so pre-occupied with our respective corners of the world that we have failed to reflect on them. Let us correct this grave error before the edifices our forebears so painstakingly built fade from the scene.

NOTES

ONE: THEORETICAL FRAMEWORK

1. Some years ago Max Black (1962) noted the important role that metaphor plays in the sciences. In an empirical study of a lab, Knorr-Cetina (1981) observed that metaphorical reasoning was central to the creation of research results. Countless others have observed the importance of metaphors in both opening and closing certain avenues of scholarship. See, for example, Hacking (1999), Lakoff and Johnson (1980), Ricoeur (1977), Turbayne (1970), and Wheelwright (1954).

2. In fairness to Hobbes (1991 [1651]), it should be noted that he was quite aware of how gender roles were created by society and not by nature.

TWO: THE CHANGING WORLD OF UNIVERSITIES

1. UC manages three Department of Energy (DOE) laboratories—each with its own ties to the Department of Defense: Los Alamos National Laboratory (LANL), Lawrence Berkeley National Laboratory (LBNL), and Lawrence Livermore National Laboratory (LLNL). R&D expenditures at the DOE labs fell more than 40 percent between 1988 and 1996 (Jaffe and Lerner 2001).

2. Universities have often been viewed as unique institutions that provide a place where scientists can carry out their work unfettered by outside interference. Like Merton's (1973) conception of science, they were often considered "above" society.

3. A striking difference between analyses of UIR and most contemporary science studies is the realist flavor of the former and the constructivist tendencies of the latter. Some of this derives from the organizational and economic sociological foundations of most UIR research, foundations that contrast with the more politicized and ethnographic traditions of technoscientific analyses. The former group focuses on the changing or problematic institutional structures of and setting for university-based scientific and technological development and the latter on the practice of university science and technological development.

THREE: LAND GRANT UNIVERSITIES, AGRICULTURAL SCIENCE, AND UC BERKELEY

1. For discussions of the history, structure, and content of shared governance at Berkeley, including a discussion of its alteration and intensification with the expansion from one to nine UC campuses, see Clark (1998) and Douglass (2002).

2. There are a few exceptions, where disciplines themselves are divided between the two kinds of research, most notably rural sociology and entomology (Buttel 1985).

3. It should be noted that the reorganization of biology at UCB was part and parcel of changes taking place in the organization of the biological sciences across all major U.S. universities at the time. The biological sciences have been largely reorganized from taxon-oriented divisions, such as zoology and botany, to divisions by level of analysis (i.e., molecule, cell, organism, ecosystem) (Roush 1997).

FOUR: A CHRONOLOGY OF EVENTS

1. The chronology is summarized in Table 4.1.

2. Important, here, is that the predominant vision embedded within these efforts was one concerned with medical and other nonplant—much less agricultural—applications.

3. That office issued guidelines that stated, "faculty members are encouraged to engage in appropriate outside professional relationships with private industry" (University of California, Office of the President 1989, 1). In addition, a UC president's retreat took place in Los Angeles, January 30–31, 1997, on university relations with industry in research and technology transfer that addressed further ways of supporting UIR.

4. It is important to note that already by 1998 the level of economic concentration in the agricultural biotechnology industry was substantial. A handful of firms, including Monsanto, Syngenta, Dow, and DuPont, dominated the fledgling agricultural biotechnology industry. Through buying seed companies and small biotechnology firms and cross-licensing patents among each other to permit freedom to operate (Barton 2002), they had already effectively established themselves as the trendsetters in the industry.

5. In 1997 DuPont purchased 20 percent of Pioneer Hi-Bred, the nation's leading producer of hybrid corn seed. It purchased the remaining shares in 1999.

6. LaPorte (2000, 67) argued that the other bids "were considered predatory and not at all in the spirit of the design; these bids were rejected, almost out of hand."

7. Novartis supports numerous "in-house" foundations. According to the company website, the Novartis Research Foundation "supports scientific research projects, particularly high-risk projects in areas of new technologies, that are compatible with the long-term interests of Novartis and its partner organizations" (Novartis 2004).

8. The UCB-N Advisory Committee was a six-member committee charged with managing the relationship between UCB and NADI, except for the specific research projects undertaken by UCB-N participants. It was made up of the vice chancellor for research, the dean of CNR, a non-CNR UCB faculty member, and three industry representatives.

9. The amendments involved minor changes in funding, waiting periods for reviews and patent claims, use of instruments provided by third parties, and the name change from NADI to TMRI. See NADI and UC (2000a, 2000b) and TMRI and UC (2002).

10. COR consisted of five members, two from NADI and three from PMB, charged with evaluating the merit of the research proposals submitted by the participating UCB-N faculty and awarding the funds.

11. Other faculty members hired by PMB during the period of UCB-N chose not to sign on to the agreement.

12. These figures only include active, pending, or issued patents. Provisional applications that were filed and then abandoned were not counted.

13. As noted above, the first survey became mired in controversy—primarily as a result of issues related to its distribution—and was deemed invalid. A second, slightly modified survey was then collected and analyzed by the California Survey Research Center—an institute affiliated with UCB—after the agreement. The surveys themselves, how they were conducted, and the findings all became the subject of further controversy. If anything, the surveys had the effect of deepening the divide between many proponents and opponents of the agreement within CNR.

14. This incident does raise a serious problem regarding the slow and deliberate character of scientific discourse and the far more rapid course of political and economic events. Critics of Quist and Chapela may well have felt that waiting for replication to take place would have resulted in unnecessarily restrictive regulation of GMOs and impeded research and commercialization.

15. The private nature of all tenure decisions is such that it would have been inappropriate for us to query participants as to the importance of Chapela's role in the controversy over UCB-N.

FIVE: POINTS OF CONTENTION

1. Graduate students and postdoctoral researchers in PMB were also largely excluded from the processes through which the agreement was developed. This will be discussed in greater depth below.

2. Such an argument skips over the likelihood of an equally dynamic result—at least on campus and within the community, if not in the press—at UC Santa Cruz, home of the Center for Agroecology and Sustainable Food Systems and the Agroecology Farm. See http://zzyx.ucsc.edu/casfs/.

3. Our figures differ slightly from the Price and Goldman report (2002, 7) because they averaged over the period 1998–2002, whereas we examined each year individually. The annual percentage of PMB's extramural funding from UCB-N continues at 26 percent (FY 2000), 39 percent (FY 2001), 20 percent (FY 2002), and 20 percent (FY 2003) (data from SPO).

SIX: OVERVIEW AND ANALYSIS OF THE AGREEMENT

1. A genome is the complete genetic code of an organism. Genomics is the study of genomes.

2. Compiled from UCB Sponsored Projects annual reports (1998, 1999, 2000, 2001, 2002, 2003).

3. See note 3 to Chapter 4, above. The president's retreat of January 30–31, 1997, further addressed ways of assisting UIR. This was of particular importance to plant genomics at the time, given the nonexistent institutional funding mechanisms within federal agencies at the time.

4. Penhoet returned to academia on July 1, 1998, as dean of UCB's School of Public Health.

5. For reviews of university-industry agreements with one firm in the 1980s, see U.S. Department of Commerce (1984), Kenney (1986), Nelkin et al. (1987), National Research Council, Committee on Japan (1992), Bowie (1994), and Martin and Reynolds (2002). See also Krimsky (2003).

6. List compiled from National Academy of Science (2003, 102–3); Reamer et al. (2003); and Slaughter and Rhoades (1996).

7. "Boundary organizations are institutions that straddle the apparent politics/science boundary and, in doing so, internalize the provisional and ambiguous character of that boundary" (Guston 2000, 30).

8. It should be noted that although most funds went to PMB, CNR negotiated the agreement, UCB's OTL wrote it, and the system-wide Board of Regents accepted it.

9. The actual range remains confidential and was not available to the external review team.

10. U.S. Patent Application 09/828,068, filed on April 6, 2001, and published electronically by the Patent and Trademark Office on October 24, 2002. A corresponding academic paper was submitted to the *Plant Cell* on March 6, 2001, accepted on May 29, 2001, and appeared in the August 2001 issue (Aubert et al. 2001).

SEVEN: THE AGREEMENT AND THE PUBLIC STAGE

1. These individuals were Thomas DiMare, former chair of the Western Growers Association and the California Tomato Board; Stuart Woolf, chair of the California League of Food Processors, and Phil Larson of the Fresno County Farm Bureau.

2. Lexis-Nexis is a database of leading newspapers and other information sources in the United States and other parts of the world.

3. A more recent follow-up search of Lexis-Nexis found only three articles on this topic, all appearing in California newspapers, between June 2002 and October 2003.

4. This article, entitled "The Kept University" (Press and Washburn 2000), used the voices heard on the Berkeley campus in the 1960s concerning ties between colleges and industry to discuss UCB-N. One of the statements, which has been echoed throughout this report, states, "that the university had the backing of a private company was hardly unusual. That a single corporation would be providing one third of the research budget of an entire department at a public university had sparked an uproar."

EIGHT: THE SCIENTIFIC ENTERPRISE

1. Numerous people, both opponents and advocates of UCB-N, argued that any effort to treat the agreement as an experiment required an external team to collect information from UCB-N's inception or, indeed, beforehand in order to have control baseline data for comparison. We first conducted interviews in January 2002, although we were able to use transcripts of some interviews that Jean Lave conducted from July to December 1999.

2. This figure is adjusted for the CRADA reduction of $1,056,000, bringing the total amount of UCB-N funds to $23,944,000 over five years.

3. Other faculty members hired by PMB during the period of UCB-N chose not to sign the agreement.

4. In academic year 2003–4, 38.5 percent of students in the graduate group in microbiology had advisors outside PMB.

5. As is the case with PMB, MSU's department is the result of a reorganization of the biological sciences in recent years.

6. The American Council on Education reported that two-thirds of public research universities made substantial program cuts in the early 1990s (Barrow 1996).

NINE: INTELLECTUAL PROPERTY RIGHTS

1. For a detailed discussion of the growth of intellectual property rights in plants, see Busch et al. (1995).

2. See Gaisford et al. (2001) for a discussion of intellectual property rents.

3. By May 2004 UCB had received option agreement fees of $60,000 from Novartis and patent cost reimbursements totaling $67,995 for patent filings, of which PMB received $5,397.

4. U.S. Patent No. 6,677,422, issued on January 13, 2004, "Stabilization of Hypoallergenic, Hyperdigestible Previously Reduced Proteins," by Bob B. Buchanan, Susumu Morigasaki, Gregorio del Val, and Oscar L. Frick, filed February 7, 2001. See also U.S. Patent No. 6,555,116,

issued on April 29, 2003, "Alleviation of the Allergenic Potential of Airborne and Contact Allergens by Thioredoxin," by Bob B. Buchanan, Gregorio del Val, Rosa M. Lozano, Joshua H. Wong, Boihon C. Yee, and Oscar L. Frick, filed on January 27, 1999 (the patent relates to patent applications dating back to October 12, 1991), and U.S. Patent Applications, publication numbers 20020098277 and 20030170763.

TEN: THE IMPACT AND SIGNIFICANCE OF UCB-N ON UCB AND CNR

1. It should be noted that acrimonious debates on this issue have been prominent in biology for some time. At Yale University such debates resulted in the division of the biology department into two distinct departments (Roush 1997).

2. Anticipating one of our conclusions in Chapter 12, such an appeal fails to understand that the tomato harvester remains near the surface in the minds of opponents as a foundational moment in the trajectories ending in UCB-N. For opponents, that proponents can make such straightforward, factual appeals is an indication of the normative problems on campus.

ELEVEN: RETHINKING THE ROLE OF PUBLIC AND LAND GRANT UNIVERSITIES

1. As is common knowledge, from the free speech movement to the nuclear weapons work at Lawrence Livermore Laboratories, UC Berkeley, has long been situated at the intersection of these diverse trends in intellectual and public life.

2. By comparison, in *The Uses of the University*, Clark Kerr (1963) divided the university into "the Platonic academy devoted to knowledge for its own sake and inspiring students to a life of inquiry; the Sophists (so despised by Plato) who aimed to impart skills useful for worldly public action; and the Baconian vision of a state-sponsored research institute devoted to producing the sort of knowledge that would extend man's dominion over nature and augment the power of the state" (Shapin 2003, 16).

3. In a brief analysis of the Novartis agreement, Krimsky (2003) likens the problems it poses to the oil leak in 1969. At that time the California attorney general was unable to find petroleum engineers at UCB or UC Santa Barbara willing to testify for fear of losing industry grants and consulting contracts.

4. Conflict of interest is typically an issue of concern where certain funds go above a trivial amount, or a *non de minimus* level, to use a common legal term. A *de minimus* calculation may be made in the case of UCB-N. The popular media rarely put university revenues from IPR into context relative to the huge scale of university finances. The impression is given that a return of millions of dollars is sure to endanger the research agenda or creditability of the institution. This is generally not the case. The Association of American Universities (AAU) estimated that in 2000 colleges and universities in the United States conducted $30 billion in research and development. The sum equals 11 percent of the nation's total of $265 billion. Academic institutions accounted for 42 percent of basic research, 13 percent of applied, and less than 1 percent of development. These institutions self-financed $6 billion of academic R&D. In 1999 UCB received $432 million in grants and contracts (NASULGC 2001, 25). Therefore, the $3.2 million received in royalties and fees (with a net income of $831,000) and the $5 million per year received from research from the Novartis (Syngenta) agreement amounted to 0.7 percent and 1.2 percent of the total, respectively. These figures appear to be *de minimis* at the campus level for the agreement and for all technology transfer activities. The same may not be true for faculty members, who could see royalty payments that are several times larger than their annual salaries. Finally, figures at the department level are reported to be between 30 and 40 percent, which is arguably *non de minimis*, causing UCB to take special precautions.

5. Nonfinancial conflicts of interest include career advancement, publications, and fame (Cicero 2003). According to the university policy on integrity in research (University of California, Office of the President 1990), misconduct in science relates to inappropriate behavior in "proposing, carrying out, or reporting results," but does not include "honest differences in interpretations or judgments of data." Similarly, the AAU defines the falsification of data as including both outright fabrication and "deceptively selective reporting" (Association of American Universities, National Association of State Universities and the Council of Graduate Schools 1993, 234).

TWELVE: CONSTRUCTING THE FUTURE

1. The episode and its aftermath may be reassembled from Jackson et al. (1972), Wade (1972), Morrow and Berg (1972), Galambos and Sewell (1995), Strickler (2001), and Stratton et al. (2002).

2. This point is developed convincingly in Kleinman and Vallas (2001). For a related defense of *basic* science, see Berg and Singer (1998) and Dasgupta and David (1994).

3. Shanmugam and Valentine (1975, 924). See Rogers (1970), Streicher et al. (1971), and Barton and Brill (1983) for other early goal statements.

4. NSF Proposal Number 9975718 (letter of intent deadline of December 4, 1998, and closing date of January 29, 1999) for Program Announcement 99-13, entitled "Global Expression Studies of the Arabidopsis Genome," p. C-1. For project outcomes, see Yamada et al. (2003).

5. For a look at standard practices, see Gunsalus and Rowan (1989). For questions on the utility of financial disclosures, see Cain et al. (2005).

6. For an overview of conflicts of interest in biomedical research, see Bekelman et al. (2003). Price (1979) explores the institutional dimension.

7. The approach is similar to the institutional conflict-of-interest recommendations for human subjects (U.S. Department of Health and Human Services 2004). However, the guidance does not include institutional management of financial conflicts to be found among investigators or nonfinancial conflicts, and is limited to research using human subjects (where the final authorization comes through an informed consent signed by the experimental subject).

8. Special interest groups include a variety of public interest associations in which administrators and investigators participate (Sonnert 2002).

9. Research funding is also in short supply in the area of environmental confinement methods for GMOs (National Research Council, Committee on the Biological Confinement of Genetically Engineered Organisms 2004).

REFERENCES

Abate, T. 2003. "Critic of Biotech Corn Fears UC Won't Give Him Tenure." *San Francisco Chronicle*, March 23, A1.

Ader, C. R. 1995. "A Longitudinal Study of Agenda Setting for the Issue of Environmental Pollution." *Journalism and Mass Communication Quarterly* 72 (2): 300–311.

Aglietta, M. 2000 [1979]. *A Theory of Capitalist Regulation: The U.S. Experience.* New York: Verso.

Allen, P. 1994. *The Human Face of Sustainable Agriculture: Adding People to the Environmental Agenda.* Santa Cruz, Calif.: Center for Agroecology and Sustainable Food Systems.

Allen, P., and C. Sachs. 1991. *What Do We Want to Sustain? Developing a Comprehensive Vision of Sustainable Agriculture.* Santa Cruz, Calif.: Center for Agroecology and Sustainable Food Systems.

Allen, P., and D. Van Dusen. 1990. *Raising Fundamental Issues.* Santa Cruz, Calif.: Center for Agroecology and Sustainable Food Systems.

Allen, P., D. Van Dusen, J. Lundy, and S. Gliessman. 1991. *Expanding the Definition of Sustainable Agriculture.* Santa Cruz, Calif.: Center for Agroecology and Sustainable Food Systems.

American Association of University Professors. 1990 [1915]. "Appendix A: General Report of the Committee on Academic Freedom and Academic Tenure (1915)." *Law and Contemporary Problems* 53: 393–406.

———. 2005a. "1940 Statement of Principles on Academic Freedom and Tenure, with 1970 Interpretive Comments." Washington, D.C.: AAUP. http://www.aaup.org/statements/Redbook/1940stat.htm (accessed December 15, 2005).

———. 2005b. "Statement on Corporate Funding of Academic Research." Washington, D.C.: AAUP. http://www.aaup.org/statements/Redbook/repcorf.htm (accessed July 1, 2005).

American Council on Education and the National Alliance of Business. 2001. "Working Together, Creating Knowledge: The University-Industry Research Collaboration." Washington, D.C.: American Council on Education and the National Alliance of Business.

American Federation of Teachers. 2005. "Academic Freedom in Higher Education." Washington, D.C.: American Federation of Teachers. http://www.aft.org/topics/academic-freedom/index.htm (accessed July 1, 2005).

Amundson, R. 2002. "Committee on Academic Freedom: Response to 'The Novartis Agreement: An Appraisal.'" Berkeley: University of California, Berkeley, Academic Senate.

Apple, R. D. 1989. "Patenting University Research: Harry Steenbock and the Wisconsin Alumni Research Foundation." *Isis* 80 (3): 374–94.

Appleseed, Inc. 2003. "Engines of Economic Growth: The Economic Impact of Boston's Eight Research Universities on the Metropolitan Boston Area." New York: Appleseed, Inc.

Arzoomanian, G. 2001. Letter to the editor. *Chronicle of Higher Education*, July 27, B18.

Association of American Universities. 2001. "Report on Individual and Institutional Financial Conflict of Interest." Washington, D.C.: Association of American Universities Task Force on Research Accountability.

———. National Association of State Universities and the Council of Graduate Schools. 1993. "Framework for Institutional Policies and Procedures to Deal with Fraud in Research." In *Responsible Science*, vol. 2, *Background Papers and Resource Documents*, ed. Panel on Scientific Responsibility and the Conduct of Research, Committee on Science, Engineering and Public Policy, National Academies of Sciences, 231–42. Washington, D.C.: National Academies Press.

Association of University Technology Managers, Inc. (AUTM). 1995–2003. "AUTM Licensing Survey: A Survey Summary of Technology Licensing (and Related) Performance for U.S. and Canadian Academic and Nonprofit Institutions, and Patent Management Firms." Northbrook, Ill.: Association of University Technology Managers, Inc.

Assouline, G., P.-B. Joly, and S. Lemarié. 2002. "Plant Biotechnology and Agricultural Supply Industry Restructuring." *International Journal of Biotechnology* 4 (2–3): 194–210.

Atkinson, R. C. 1999. "The Future Arrives First in California." *Issues in Science and Technology* 16 (1): 45–51.

Aubert, D., L. Chen, Y. H. Moon, D. Martin, L. A. Castle, C. H. Yang, and Z. R. Sung. 2001. "EMF1, a Novel Protein Involved in the Control of Shoot Architecture and Flowering in Arabidopsis." *Plant Cell* 13 (8): 1865–75.

Bacon, F. 1994 [1620]. *Novum Organum*. Chicago: Open Court.

Bagdikian, B. 1992. *The Media Monopoly*. 4th ed. Boston: Beacon Press.

Barinaga, M. 1994. "A Bold New Program at Berkeley Runs into Trouble." *Science* 263 (5152): 1367–68.

Barnes, M., and P. S. Florencio. 2002. "Investigator, IRB and Institutional Financial Conflicts of Interest in Human-Subjects Research: Past, Present and Future." *Seton Hall Law Review* 35: 525–61.

Barrow, C. W. 1996. "The Strategy of Selective Excellence: Redesigning Higher Education for Global Competition in a Postindustrial Society." *Higher Education* 31 (4): 447–69.

Barton, J. H. 2002. "Antitrust Treatment of Oligopolies with Mutually Blocking Patent Portfolios." *Antitrust Law Journal* 69 (3): 851–83.

Barton, K., and W. Brill. 1983. "Prospects in Plant Genetic Engineering." *Science* 219 (4585): 671–76.

Bayh, B. 2004. "Statement of Senator Birch Bayh to the National Institutes of Health, May 25, 2004." Available at http://ott.od.nih.gov/Meeting/Senator-Birch-Bayh.pdf.

Becher, T., and P. R. Trowler. 2001. *Academic Tribes and Territories*. 2d ed. Buckingham, UK: Society for Research in Higher Education.

Beck, U. 1992. *Risk Society: Towards a New Modernity*. London: Sage Publications.

———. 1999. *World Risk Society*. Malden, Mass.: Polity Press.

Beck, U., A. Giddens, and S. Lash. 1994. *Reflexive Modernization: Politics, Tradition and Aesthetics in the Modern Social Order*. Cambridge, UK: Polity Press.

Bekelman, J., Y. Li, and C. Gross. 2003. "Scope and Impact of Financial Conflicts of Interest in Biomedical Research." *JAMA* 289 (4): 454–65.

Bella, D. A. 1985. "The University: Eisenhower's Warning Reconsidered." *Journal of Professional Issues in Engineering* 111 (1): 12–21.

Benfey, O., and P. Morris, eds. 2001. *Robert Burns Woodward: Architect and Artist in the World of Molecules*. Philadelphia: Chemical Heritage Foundation.

Benjamin, M. 1990. *Splitting the Difference: Compromise and Integrity in Ethics and Politics*. Lawrence: University Press of Kansas.

Bennett, W. J. 1992. *The De-valuing of America: The Fight for Our Culture and Our Children*. New York: Summit Books.

Berg, P., and M. Singer. 1998. "Inspired Choices." *Science* 282 (5390): 873–74.

Berneman, L. P. 1995. "Inside Industry, University Licensing." *Les Nouvelles* 30 (3): 134–35.

Besse, I., J. H. Wong, K. Kobrehel, and B. B. Buchanan. 1996. "Thiocalsin: A Thioredoxin-Linked, Substrate-Specific Protease Dependent on Calcium." *Proceedings of the National Academy of Sciences* 93 (8): 3169–75.

Best, J. 1991. "'Road Warriors' on 'Hair-Trigger Highways': Cultural Resources and the Media's Construction of the 1987 Freeway Shootings Problem." *Sociological Inquiry* 61 (3): 327–45.

Beus, C. E., and R. E. Dunlap. 1990. "Conventional versus Alternative Agriculture: The Paradigmatic Roots of the Debate." *Rural Sociology* 55 (4): 590–616.

Bibel, D. 1995. "For the Welfare of the Nation: Three Stories of Medical Microbiology at the University of California, History Lecture for the Sixtieth Anniversary of the Northern California Branch of the American Society for Microbiology." In *Northern California Branch of the American Society for Microbiology: Sixtieth Anniversary, 1935–1995 Commemorative Edition*, ed. Archives Committee, 10–14. Oakland, Calif.: American Society for Microbiology, Northern California Branch.

Biggart, N., and T. Beamish. 2003. "The Economic Sociology of Conventions: Habit, Custom, Practice, and Routine in Market Order." *Annual Review of Sociology* 29: 443–64.

Billings, J. 1886. "Scientific Men and Their Duties." *Science* 8: 541–51.

Bird, E. A. R. 1993. "Politics of Nature: An Interpretation of the Environmental Release Controversy Featuring Genetically Engineered Crop Protection." PhD diss., University of California, Santa Cruz, History of Consciousness Department.

Black, M. 1962. *Models and Metaphors: Studies in Language and Philosophy*. Ithaca: Cornell University Press.

Bloom, A. D. 1987. *The Closing of the American Mind: How Higher Education Has Failed Democracy and Impoverished the Souls of Today's Students*. New York: Simon and Schuster.

Bloor, D. 1991. *Knowledge and Social Imagery*. Chicago: University of Chicago Press.

Blout, E. 2001. "Robert Burns Woodward: April 10, 1917–July 8, 1979." *Biographical Memoirs (National Academy of Sciences)* 80: 366–87.

Blumenstyk, G. 1998. "Berkeley Pact with a Swiss Company Takes Technology Transfer to a New Level." *Chronicle of Higher Education*, December 11, A56–57.

———. 2003. "The Price of Research." *Chronicle of Higher Education*, December 19, A26.

Blumenthal, D., M. Gluck, K. S. Louis, M. A. Stoto, and D. Wise. 1986a. "Industry Research Relationships in Biotechnology: Implications for the University." *Science* 232 (4756): 1361–66.

Blumenthal, D., M. Gluck, K. S. Louis, and D. Wise. 1986b. "Industrial Support of University Research in Biotechnology." *Science* 231 (4735): 242–46.

Blumer, H. 1971. "Social Problems as Collective Behavior." *Social Problems* 18 (3): 298–305.

Bok, D. 2003. *Universities in the Marketplace: The Commercialization of Higher Education*. Princeton: Princeton University Press.

Boltanski, L., and L. Thévenot. 1991. *De La Justification: Les Economies de la Grandeur*. Paris: Gallimard.

———. 1999. "The Sociology of Critical Capacity." *European Journal of Social Theory* 2 (3): 359–77.

Bourdieu, P. 1997. *Les Usages Sociaux de la Science*. Paris: Institute National de la Recherche Agronomique.

Bowie, N. E. 1994. *University-Business Partnership: An Assessment*. Lanham, Md.: Rowman & Littlefield.

Boyer, R., and Y. Saillard. 2002. *Régulation Theory: The State of the Art*. London: Routledge.

Brannigan, A. 1981. *The Social Basis of Scientific Discoveries*. Cambridge: Cambridge University Press.

Bremer, H. W. 2001. "The First Two Decades of the Bayh-Dole Act as Public Policy." Washington, D.C.: National Association of State Universities and Land-Grant Colleges.

Brentano, R. 1998a. Letter to Professor David Littlejohn, Chair, Academic Freedom Committee, August 26. University of California, Berkeley. In the archives of the UC Berkeley Academic Senate.

———. 1998b. Letter to Carol Christ, Executive Vice Chancellor and Provost, October 6. University of California, Berkeley. In the archives of the UC Berkeley Academic Senate.

———. 1998c. Letter to Carol Christ, Executive Vice Chancellor and Provost, November 18. University of California, Berkeley. In the archives of the UC Berkeley Academic Senate.

Brock, W. 2000. *The Chemical Tree: A History of Chemistry*. New York: W. W. Norton.

Brody, H. 2001. "The TR University Research Scorecard." *Technology Review* 104 (September): 81–83.

Brooks, H., and L. Randazzese. 1999. "University-Industry Relations: The Next Four Years and Beyond." In *Investing in Innovation: Creating a Research and Innovation Policy*, ed. L. Branscomb and J. Keller, 361–99. Cambridge: MIT Press.

Bucchini, L., and L. Goldman. 2002. "Starlink Corn: A Risk Analysis." *Environmental Health Perspectives* 110 (1): 5–13.

Buchanan, B. B. 2001. "Genetic Engineering and the Allergy Issue." *Plant Physiology* 126 (11): 5–7.

Buchanan, B. B., C. Adamidi, R. M. Lozano, B. C. Yee, M. Momma, K. Kobrehel, R. Ermel, and O. L. Frick. 1997. "Thioredoxin-Linked Mitigation of Allergic Responses to Wheat." *Science* 94 (10): 5372–77.

Burress, C. 2005. "Professor Who Was Denied Tenure Sues UC." *San Francisco Chronicle*, April 19, B3.

Busch, L., and W. B. Lacy. 1983. *Science, Agriculture, and the Politics of Research*. Boulder, Colo.: Westview Press.

Busch, L., W. B. Lacy, J. Burkhardt, D. Hemken, J. Moraga-Rojel, J. D. Souza Silva, and T. Koponen. 1995. *Making Nature, Shaping Culture: Plant Biodiversity in Global Context*. Lincoln: University of Nebraska Press.

Busch, N. A. 2002. "Jack and the Beanstalk: Property Rights in Genetically Modified Plants." *Minnesota Intellectual Property Review* 3 (2): 1–235. Available at http://mipr.umn.edu/archive/v3n2/busch.pdf.

Bush, V. 1945. *Science: The Endless Frontier*. Washington, D.C.: U.S. Government Printing Office.

Buttel, F. H. 1985. "The Land-Grant System: A Sociological Perspective on Value Conflicts and Ethical Issues." *Agriculture and Human Values* 2 (2): 78–95.

———. 1986. "Biotechnology and Public Agricultural Research Policy: Emergent Issues." In *New Directions for Agriculture and Agricultural Research: Neglected Dimensions and Emerging Alternatives*, ed. K. Dahlberg, 312–47. Totowa, N.J.: Rowman & Allanheld.

Cain, D., G. Loewenstein, and D. Moore. 2005. "Coming Clean But Playing Dirtier: The Shortcomings of Disclosure as a Solution to Conflicts of Interest." In *Conflicts of Interest*, ed. D. Moore, D. Cain, G. Loewenstein, and M. Bazerman, 104–25. New York: Cambridge University Press.

California Council on Science and Technology. 2002. "Benefits and Risks of Food Biotechnology." Sacramento, Calif.: California Council on Science and Technology.

California Senate Committee on Natural Resources and Wildlife, California Senate Select Committee on Higher Education. 2000. "Impacts of Genetic Engineering on California's

Environment: Examining the Role of Research at Public Universities (Novartis/UC Berkeley Agreement)." Sacramento: Senate Publications.

Callon, M. 1986. "Some Elements of a Sociology of Translation: Domestication of the Scallops and the Fishermen of St. Brieuc Bay." In *Power, Action and Belief: A New Sociology of Knowledge?* ed. J. Law, 196–233. London: Routledge & Kegan Paul.

———. 1998. "Introduction: The Embeddedness of Economic Markets in Economics." In *The Laws of the Market*, ed. M. Callon, 1–57. Oxford: Blackwell.

Calvert, J. 2004. "The Idea of 'Basic Research' in Language and Practice." *Minerva* 42 (3): 251–68.

Calvert, J., and B. Martin. 2001. *Changing Conceptions of Basic Research?* Background document prepared for the OECD/Norwegian Workshop on Policy Relevance and Measurement of Basic Research, Oslo, Norway, October 29–30, 2001. Brighton, UK: Science and Technology Policy Research, University of Sussex. Available at https://www.oecd.org/ dataoecd/39/0/ 2674369.pdf.

Campbell, P. 2002. "Editorial Note." *Nature* 16 (April 4): 600.

Cash, D. 2001. "'In Order to Aid in Diffusing Useful and Practical Information': Agricultural Extension and Boundary Organizations." *Science, Technology and Human Values* 26 (4): 431–52.

Castells, M. 1991. *The Informational City: Information, Technology, Economic Restructuring, and the Urban/Regional Process*. Oxford: Basil Blackwell.

———. 1996. *The Rise of the Network Society*. Cambridge, Mass.: Blackwell Publishers.

Cavanaugh, D. C. 1974. "Karl Friedrich Meyer." *Journal of Infectious Diseases* 129 (supplement): 3–10.

Charles, D. 2001. *Lords of the Harvest: Biotech, Big Money, and the Future of Food*. Cambridge, Mass.: Perseus Publishing.

Cho, M. J., J. H. Wong, C. Marx, W. Jiang, P. G. Lemaux, and B. B. Buchanan. 1999. "Overexpression of Thioredoxin h Leads to Enhanced Activity of Starch Debranching Enzyme (Pullulanase) in Barley Grain." *Proceedings of the National Academy of Sciences* 96 (25): 14641–46.

Christ, C. T. 1998a. Letter to Robert J. Brentano, answers to DIVCO, October 16, University of California, Berkeley. In the archives of the UC Berkeley Academic Senate.

———. 1998b. Letter to Robert J. Brentano, Chair, Academic Senate, November 22. University of California, Berkeley. In the archives of the UC Berkeley Academic Senate.

———. 1999. "Answers to Questions Posed by Members of the Faculty and Committees of the Academic Senate on the UC Berkeley–Novartis Research Agreement." February 17 (revised October 16, 1999). University of California, Berkeley. In the archives of the UC Berkeley Academic Senate.

———. 2000. Letter to Ms. Adkins, May 2. University of California, Berkeley. In the archives of the UC Berkeley Academic Senate.

Cicero, T. J. 2003. "Conflict of Interest Issues and Trends." Paper presented at the AUTM Annual Meeting, February 7, Orlando, Florida.

Clark, B. 1997. "Small Worlds, Different Worlds: The Uniquenesses and Troubles of American Academic Professions." *Daedalus* 126 (4): 21–42.

———. 1998. *Creating Entrepreneurial Universities: Organizational Pathways of Transformation*. New York: Pergamon, for IAU Press.

Clarke, A. 1997. "A Social World's Research Adventure: The Case of Reproductive Science." In *Grounded Theory in Practice*, ed. A. Strauss and J. Corbin, 63–94. Thousand Oaks, Calif.: Sage Publications.

Clarke, J., and J. Newman. 1997. *The Managerial State: Power, Politics, and Ideology in the Remaking of Social Welfare*. London: Sage Publications.

Cochrane, R. 1947. *History of the Chemical Warfare Service in World War II (1 July 1940–15 August 1945), Biological Warfare Research in the United States*. Army Chemical Center, Md.: Chemical Corps.

Collier, J. 2002. "Scripting the Radical Critique of Science: The Morill Act and the American Land-Grant University." *Futures* 34 (2): 182–91.

Condit, P., and R. B. Pipes. 1997. "The Global University." *Issues in Science and Technology* 14: 27–28.

Committee to Review Swedish Research Policy. 1998. "Slutbetänkande av Kommittén för översyn av den svenska sorskningspolitiken, Final Report." Stockholm: Institute for Labor Market Policy Evaluation.

Cortwright, J., and H. Mayer. 2002. *Signs of Life: The Growth of Biotechnology Centers in the U.S.* Washington D.C.: Brookings Institute.

Council on Government Relations. 1993. "University Technology: Questions and Answers." Washington, D.C.: Council on Government Relations.

Creager, A. 2002. *The Life of a Virus: Tobacco Mosaic Virus as an Experimental Model, 1930–1965.* Chicago: University of Chicago Press.

Crouch, M. L. 1990. "Debating the Responsibilities of Plant Scientists in the Decade of the Environment." *Plant Cell* 2 (4): 275–77.

Culliton, B. J. 1977. "Harvard and Monsanto: The $23-Million Alliance." *Science* 195 (4280): 759–63.

Cummings, J., and B. Teng. 2003. "Transferring R&D Knowledge: The Key Factors Affecting Knowledge Transfer Success." *Journal of Engineering and Technology Management* 20 (1–2): 39–68.

Daniels, G. H. 1967. "The Pure Science Ideal and Democratic Culture." *Science* 156 (3783): 1699–1705.

Dasgupta, P., and P. David. 1994. "Toward a New Economics of Science." *Research Policy* 23 (5): 487–521.

Davidson, P. W. 1983. "The Third-Person Effect in Communication." *Public Opinion* 47 (1): 1–15.

Day, B. A. 1978. "The Morality of Agronomy." In *Agronomy in Today's Society*, ed. J. W. Pendleton, 19–27. ASA Special Publication 33. Washington D.C.: American Sociological Association.

Dearing, J. W., and E. M. Rogers. 1996. *Agenda Setting.* Thousand Oaks, Calif.: Sage Publications.

Delanty, G. 2001a. *Challenging Knowledge: The University in the Knowledge Society.* Philadelphia: Society for Research into Higher Education.

———. 2001b. "The University in the Knowledge Society." *Organization* 8 (2): 149–53.

Deutsche Bank. 1999. "Ag Biotech: Thanks, But No Thanks?" New York: Deutsche Bank.

Douglass, J. A. 2002. "From Multi- to Meta-University: Organizational and Political Change at the University of California in the 20th Century and Beyond." CSHE Paper 4-02. Berkeley: University of California, Berkeley, Center for the Study of Higher Education.

Durham, F. 2005. "Public Relations as Structuration: A Prescriptive Critique of the StarLink Global Food Contamination Case." *Journal of Public Relations Research* 17 (1): 29–47.

Dworkin, R. 1996. "We Need a New Interpretation of Academic Freedom." In *The Future of Academic Freedom*, ed. L. Menand, 181–98. Chicago: University of Chicago Press.

Edsall, J. T. 1975. "Scientific Freedom and Responsibility." *Science* 188 (4189): 687–93.

Eisenhower, D. D. 1961. "Farewell Address." In *Public Papers of the Presidents, Dwight D. Eisenhower, Containing the Public Messages, Speeches and Statements of the President, 1953–1960/61.* Vol. 1, 1035–40. Washington, D.C.: U.S. Government Printing Office.

Eisinger, P. 1988. *The Rise of the Entrepreneurial State.* Madison: University of Wisconsin Press.

Ericson, R., P. Baranek, and J. Chan. 1989. *Negotiating Control.* Toronto: University of Toronto Press.

Eschenmoser, A. 2001. "RBW, Vitamin B_{12}, and the Harvard-ETH Collaboration." In *Robert Burns Woodward: Architect and Artist in the World of Molecules*, ed. O. Benfey and P. Morris, 23–38. Philadelphia: Chemical Heritage Foundation.

Etzkowitz, H. 1998. "The Norms of Entrepreneurial Science: Cognitive Effects of the New University-Industry Linkages." *Research Policy* 27 (8): 823–33.

———. 2003. "Innovation in Innovation: The Triple Helix of University-Industry-Government Relations." *Social Science Information* 42 (3): 293–337.

Etzkowitz, H., and L. Leydesdorff. 1997. *Universities and the Global Knowledge Economy: A Triple Helix of University-Industry-Government Relations.* New York: Pinter.

———. 1998. "The Endless Transition: A 'Triple Helix' of University-Industry-Government Relations." *Minerva* 36 (3): 203–8.

Etzkowitz, H., A. Webster, and P. Healey. 1998. "Introduction." In *Capitalizing Knowledge: New Intersections of Industry and Academia,* ed. A. Webster, P. Healey, and H. Etzkowitz, 1–17. Albany: State University of New York Press.

Fairweather, J. S. 1988. *Entrepreneurship and Higher Education: Lessons for Colleges, Universities, and Industry.* ASHE-ERIC Higher Education Report No. 6. Washington, D.C.: Association for the Study of Higher Education.

Fejer, S. O. 1966. "The Problem of Plant Breeders' Rights." *Agricultural Science Review* 4 (3): 1–7.

Feller, I. 1987. "Technology Transfer, Public Policy, and the Cooperative Extension Service-OMB Imbroglio." *Journal of Policy Analysis and Management* 6 (3): 307–27.

Feller, I., P. Madden, L. Kaltreider, D. Moore, and L. Sims. 1987. "The New Agricultural Research and Technology Transfer Policy Agenda." *Research Policy* 16: 315–25.

Fiske, J. 1992. "Audiencing: A Cultural Studies Approach to Watching Television." *Poetics* 21 (4): 345–59.

Fox, J. L. 1981. "Can Academia Adapt to Biotechnology's Lure?" *Chemical and Engineering News* 59 (41): 39–44.

Franklin, B., and D. Murphy, eds. 1998. *Making the Local News.* New York: Routledge.

Frey, K. 1996. "National Plant Breeding Study I: Human and Financial Resources Devoted to Plant Breeding Research and Development in the United States in 1994." Ames: Iowa Agriculture and Home Economics Experiment Station.

Friedland, W. H., and A. Barton. 1975. *Destalking the Wily Tomato.* Research Monograph No. 15. Davis: University of California, Davis, Department of Applied Behavioral Sciences.

Fuglie, K., N. Ballenger, K. Day, C. Ollinger, M. Reilly, J. Vassavada, and J. Yee. 1996. "Agricultural Research and Development: Public and Private Investments under Alternative Markets and Institutions." USDA/ERS, Agricultural Economics Report No. 735. Washington, D.C.: USDA/ERS.

Fujimura, J. 1988. "The Molecular Biological Bandwagon in Cancer Research: Where Social Worlds Meet." *Social Problems* 35 (3): 261–83.

Fukuyama, F. 2002. *Our Posthuman Future.* New York: Farrar, Straus and Giroux.

Fuller, S. 2000. *The Governance of Science: Ideology and the Future of the Open Society.* Philadelphia: Open University Press.

Gaisford, J., J. Hobbs, W. Kerr, N. Perdikis, and M. Plunkett. 2001. *The Economics of Biotechnology.* Northampton, Mass.: Edward Elgar.

Galambos, L., and J. Sewell. 1995. *Networks of Innovation: Vaccine Development at Merck, Sharp & Dohme, and Mulford, 1895–1995.* New York: Cambridge University Press.

Gamson, W. A. 1984. *What's News: A Game Simulation of TV News.* New York: Free Press.

Gamson, W. A., D. Croteau, W. Hoynes, and T. Sasson. 1992. "Media Images and the Social Construction of Reality." *Annual Review of Sociology* 18: 373–93.

Gamson, W. A., and A. Modigliani. 1989. "Media Discourse and Public Opinion on Nuclear Power: A Constructionist Approach." *American Journal of Sociology* 95 (1): 1–37.

Gans, H. 1979. *Deciding What's News.* New York: Vintage.

Gaskell, G., and M. W. Bauer, eds. 2001. *Biotechnology, 1996–2000.* London: Science Museum.

Geiger, R. 2004. *Knowledge and Money: Research Universities and the Paradox of the Marketplace.* Stanford: Stanford University Press.

George, A., and A. Bennett. 2005. *Case Studies and Theory Development in the Social Sciences.* Cambridge: MIT Press.

Giri, A. 2000. "Audited Accountability and the Imperative of Responsibility." In *Audit Cultures: Anthropological Studies in Accountability, Ethics, and the Academy*, ed. M. Strathern, 173–95. New York: Routledge.

Gitlin, T. 1980. *The Whole World Is Watching*. Berkeley and Los Angeles: University of California Press.

Glass, B. 1965. *Science and Ethical Values*. Chapel Hill: University of North Carolina Press.

Goldschmidt, W. R. 1978 [1946]. *As You Sow: Three Studies in the Social Consequences of Agribusiness*. Montclair, N.J.: Allanheld Osmun.

Gray, P. 2001. Letter to Mr. Hesterman, February 2. University of California, Berkeley. In the archives of the UC Berkeley Academic Senate.

Greenberg, D. S. 1966. "Bootlegging: It Holds a Firm Place in Conduct of Research." *Science* 153: 848–49.

———. 2001. *Science, Money, and Politics: Political Triumph and Ethical Erosion*. Chicago: University of Chicago Press.

Gross, P. R., and N. Levitt. 1998. *The Higher Superstition: The Academic Left and Its Quarrels with Science*. Baltimore: Johns Hopkins University Press.

Gruissem, W. 1997. Letter to the International Biotechnology Advisory Board Members, March 21. University of California, Berkley. In the archives of the UC Berkeley Academic Senate.

Gunsalus, C., and J. Rowan. 1989. "Conflict of Interest in the University Setting: I Know It When I See It." *Research Management Review* 3: 13–26.

Gunther, A. 1992. "Biased Press or Biased Public? Attitudes toward Media Coverage of Social Groups." *Public Opinion Quarterly* 56 (2): 147–67.

Guston, D. 2000. *Between Politics and Science: Assuring the Integrity and Productivity of Research*. Cambridge: Cambridge University Press.

Guston, D., W. Clark, T. Keating, D. Cash, S. Moser, C. Miller, and C. Powers. 2000. "Report of the Workshop on Boundary Organizations in Environmental Policy and Science Workshop." Cambridge: Harvard University, John F. Kennedy School of Government, Environment and Natural Resources Program, Global Environmental Assessment Project.

Hackett, E. 1990. "Science as a Vocation in the 1990s: The Changing Organizational Culture of Academic Science." *Journal of Higher Education* 61 (3): 241–79. Reprinted in *Degrees of Compromise: Industrial Interests and Academic Values*, ed. J. Croissant and S. Restivo (Albany: State University of New York Press, 2001), 101–38.

Hacking, I. 1999. *The Social Construction of What?* Cambridge: Harvard University Press.

Hadwiger, D. 1982. *The Politics of Agricultural Research*. Lincoln: University of Nebraska Press.

Haraway, D. J. 1991. *Simians, Cyborgs, and Women: The Reinvention of Nature*. London: Free Association.

———. 1997. *Modest_Witness@Second_Millennium.FemaleMan©Meets OncoMouse™*. New York: Routledge.

Hardin, C. M. 1955. *Freedom in Agricultural Education*. Chicago: University of Chicago Press.

Harding, S. 1991. *Whose Science? Whose Knowledge? Thinking from Women's Lives*. Ithaca: Cornell University Press.

Hartman, H., M. Wu, B. B. Buchanan, and J. C. Gerhart. 1993. "Spinach Thioredoxin m Inhibits DNA Synthesis in Fertilized Xenopus Eggs." *Proceedings of the National Academy of Sciences* 90 (6): 2271–75.

Harvey, D. 1989. *The Condition of Postmodernity: An Enquiry into the Origins of Cultural Change*. New York: Blackwell.

———. 1998. "University, Inc." *Atlantic Monthly* October, 112–16.

Hassanein, N. 2000. "Democratizing Agricultural Knowledge through Sustainable Farming Networks." In *Science, Technology, and Democracy*, ed. D. Kleinman, 49–66. Albany: State University of New York Press.

Hatakenada, S. 2004. *University-Industry Partnerships in MIT, Cambridge, and Tokyo: Storytelling across Boundaries*. New York: Routledge.

Hays, S. P. 1959. *Conservation and the Gospel of Efficiency: The Progressive Conservation Movement, 1890–1920*. Cambridge: Harvard University Press.

Heller, D. 2001. "Uncertain Times: State Funding for Higher Education Shrinks and Tuitions Rise." *National CrossTalk* 9 (4): 11–12.

Hesterman, O. 2000. Letter to Dr. Carol T. Christ, September 6. W. K. Kellogg Foundation, Battle Creek, Mich. In the archives of the UC Berkeley Academic Senate.

Hightower, J. 1973. *Hard Tomatoes, Hard Times*. Cambridge, Mass.: Schenckman.

Hileman, B. 2002. "What's Hiding in Transgenic Foods?" *Chemical and Engineering News* 80 (1): 20–23.

Hilgard, E. 1882. "Progress in Agriculture by Education and Government Aid, II." *Atlantic Monthly*, May, 651–61.

Hilgartner, S., and C. Bosk. 1988. "The Rise and Fall of Social Problems: A Public Arenas Model." *American Journal of Sociology* 94 (1): 53–78.

Hiltzik, M. 2005. "California: Golden State; Biotech Deal Still Clouds Tenure." *Los Angeles Times*, July 7, C1.

Hobbes, T. 1996 [1651]. *The Leviathan*. Cambridge: Cambridge University Press.

Hoijer, B. 1992. "Reception of Television Narration as a Socio-Cognitive Process: A Schema-theoretical Outline." *Poetics* 21: 283–304.

Hounshell, D., and J. Smith Jr. 1988. *Science and Corporate Strategy: DuPont R&D, 1902–1980*. New York: Cambridge University Press.

Hoynes, W., and D. Croteau. 1991. "The Chosen Few: *Nightline* and the Politics of Public Affairs Television." *Critical Sociology* 18 (1): 19–34.

Hughes, S. S. 2001. "Making Dollars Out of DNA: The First Major Patent in Biotechnology and the Commercialization of Molecular Biology, 1974–1980." *Isis* 92 (3): 541–75.

Hull, D. 1988. *Science as a Process: An Evolutionary Account of the Social and Conceptual Development of Science*. Chicago: University of Chicago Press.

ICF Consulting. 2003. *California's Future: It Starts Here—UC's Contribution to Economic Growth, Health, and Culture, an Impact Study for the University of California*. San Francisco: ICF Consulting.

Isserman, A. M. 2000. "Mobilizing a University for Important Social Science Research: Biotechnology at the University of Illinois." *American Behavioral Scientist* 44 (3): 310–17.

Jackson, D., R. Symons, and P. Berg. 1972. "Biochemical Method for Inserting New Genetic Information into DNA of Simian Virus 40: Circular SV40 DNA Molecules Containing Lambda Phage Genes and the Galactose Operon of *Escherichia coli*." *Proceedings of the National Academy of Sciences* 69 (10): 2904–9.

Jaffe, A. B. 2000. "The U.S. Patent System in Transition: Policy Innovation and the Innovation Process." *Research Policy* 29 (4–5): 531–57.

Jaffe, A. B., and J. Lerner. 2001. "Reinventing Public R&D: Patent Policy and the Commercialization of National Laboratory Technologies." *RAND Journal of Economics* 32 (1): 167–98.

Jasanoff, S. 1990. *The Fifth Branch: Science Advisers as Policymakers*. Cambridge: Harvard University Press.

Jessop, B. 2002. *The Future of the Capitalist State*. Cambridge, UK: Polity Press.

Johns, M., M. Barnes, and P. Florencio. 2003. "Restoring Balance to Industry-Academia Relationships in an Era of Institutional Financial Conflicts of Interest: Promoting Research While Maintaining Trust." *JAMA* 289 (6): 741–46.

Kaplan, A. 1964. *The Conduct of Inquiry*. Scranton, Pa.: Chandler.

Kaplan, M. 1986. "Universities, Centers, and Economic Development: Converting Rhetoric to Reality." In *Issues in Higher Education and Economic Development*, 99–108. Washington, D.C.: American Association of State Colleges and Universities.

Kasper, C. G. 2000. "Novartis Agricultural Discovery Institute, Inc. (A)." Cambridge: Harvard Business School.

Kaufman, T., and E. Rúveda. 2005. "The Quest for Quinine: Those Who Won the Battles and Those Who Won the War." *Angewandte Chemie*, international ed., 44 (6): 854–85.

Kay, L. 1986. "W. M. Stanley's Crystallization of the Tobacco Mosaic Virus, 1930–1940." *Isis* 77 (288): 450–72.

Keller, E. 1985. *Reflections on Gender and Science*. New Haven: Yale University Press.

Kenney, M. 1986. *Biotechnology: The University-Industry Complex*. New Haven: Yale University Press.

Kerr, C. 1963. *The Uses of the University*. Cambridge: Harvard University Press.

Keyes, J. A., and K. A. Jones. 2001. "Brief for Association of American Medical Colleges et al., as Amici Curiae in Support of Petitioner, *Duke University v. John M. Madey*." Washington, D.C.: Association of American Medical Colleges.

Killoren, R. 1989. "Institutional Conflict of Interest." *Research Management Review* 3 (2): 1–12.

Kimber, I., and R. Dearman. 2001. "Approaches to Assessment of the Allergenic Potential of Novel Proteins in Food from Genetically Modified Crops." *Toxicological Sciences* 68 (1): 4–8.

Kitcher, P. 2001. *Science, Truth, and Democracy*. Oxford: Oxford University Press.

Kleinman, D. L. 2003. *Impure Cultures: University Biology and the World of Commerce*. Madison: University of Wisconsin Press.

Kleinman, D. L., and S. Vallas. 2001. "Science, Capitalism, and the Rise of the 'Knowledge Worker': The Changing Structure of Knowledge Production in the United States." *Theory and Society* 30 (4): 451–92.

Kloppenburg, J. R., Jr. 1988. *First the Seed: The Political Economy of Plant Biotechnology, 1492–2000*. Cambridge: Cambridge University Press.

Knight, J. 2003. "A Dying Breed." *Nature* 421 (February 6): 568–70.

Knorr-Cetina, K. D. 1981. *The Manufacture of Knowledge*. Oxford: Pergamon Press.

Knudson, T., and M. Lee. 2004. "Part Three: Biotech Industry Funds Bumper Crop of UC Davi Research." *Sacramento Bee*, June 8. http://www.sacbee.com/static/live/news/projects/biotech/c3_1.html.

Korn, D. 2000. "Conflicts of Interest in Biomedical Research." *JAMA* 284: 2234–37.

Kornberg, A. 1995. *The Golden Helix: Inside Biotech Ventures*. Sausalito: University Science Books.

Krimsky, S. 2003. *Science and the Private Interest*. Lanham, Md.: Rowman & Littlefield.

Kuhn, T. 1962. *The Structure of Scientific Revolutions*. Chicago: University of Chicago Press.

Lakoff, G., and M. Johnson. 1980. *Metaphors We Live By*. Chicago: University of Chicago Press.

Lange, J. I. 1993. "The Logic of Competing Information Campaigns: Conflict over Old Growth and the Spotted Owl." *Communication Monographs* 60 (3): 237–57.

LaPorte, T. R. 1999. Memo to Carol Christ, June. University of California, Berkeley. In the archives of the UC Berkeley Academic Senate.

———. 2000. "Diluting Public Patrimony or Inventive Response to Increasing Knowledge Asymmetries: Watershed for Land Grant University? Reflections on the University of California, Berkeley–Novartis Agreement." In *Research Terms and Partnerships: Trends in the Chemical Sciences*, 66–84. Washington, D.C.: National Academies Press.

Latour, B. 1987. *Science in Action*. Cambridge: Harvard University Press.

———. 1993. *We Have Never Been Modern*. Cambridge: Harvard University Press.

———. 2002. *War of the Worlds: What about Peace?* Chicago: Prickly Paradigm Press.

———. 2005. *Reassembling the Social: An Introduction to Actor-Network-Theory*. New York: Oxford University Press.

Lau, E. 2005. "UC Biotech Critic Reinstated, Given Tenure." *Sacramento Bee*, May 21, A3.

Lave, J. 1999. Letter to Chancellor Berdahl, Executive Vice-Chancellor Christ, Professor Spear, Chair Academic Senate, Professor LaPorte, Chair COR, November 8. University of California, Berkeley. In the archives of the UC Berkeley Academic Senate.

Lefkowitz, J. 2003. *Ethics and Values in Industrial-Organizational Psychology*. Mahwah, N.J.: Lawrence Erlbaum Associates.

Levin, R. 2001. "The University as an Engine of Economic Growth." Talk given at Tsinghua University, Beijing, China, May 9. http://www.yale.edu/opa/president/speeches/.

Leydesdorff, L., and H. Etzkowitz. 2001. "The Transformation of University-Industry-Government Relations." *Electronic Journal of Sociology* 5 (4). http://www.sociology.org/content/vol005.004/th.html.

Littlejohn, D. 1998. Letter to Professor Robert Brentano, October 29. University of California, Berkeley. In the archives of the UC Berkeley Academic Senate.

Luria, S. 1984. *A Slot Machine, a Broken Test Tube: An Autobiography*. New York: Harper & Row.

Macilwain, C. 2005. "U.S. Launches Probe into Sales of Unapproved Transgenic Corn." *Nature* 434 (March 23): 423.

MacLachlan, A. J. 2000. "Impact of the Novartis Agreement on Graduate Students in the Plant and Microbial Biology Department." University of California, Berkeley, Center for Studies in Higher Education. In the archives of the UC Berkeley Academic Senate.

Marcy, W. 1978. "Patent Policies at Educational and Nonprofit Scientific Institutions." In *Patent Policy: Government, Academic, and Industrial Concepts*, ed. W. Marcy, 78–89. Washington, D.C.: American Chemical Society

Marrs, B. 2002. "The Early History of the Genetics of Photosynthetic Bacteria: A Personal Account." *Photosynthesis Research* 73 (1–3): 55–58.

Marshall, J. 1997. "Group of UC Economists Hope to Cash in on IPO." *San Francisco Chronicle*, October 30, C1.

Martin, J., and T. Reynolds. 2002. "Academic-Industrial Relationships: Opportunities and Pitfalls." *Science and Engineering Ethics* 8 (3): 443–54.

McClung, L., and K. Meyer. 1974. "Beginnings of Bacteriology in California." *Bacteriological Reviews* 38 (3): 251–71.

McConnell, G. 1953. *The Decline of Agrarian Democracy*. Berkeley and Los Angeles: University of California Press.

Mena, J., and R. Sanders. 1998. "Swiss Pharmaceutical Company Novartis Commits $25 Million to Support Biotechnology Research at UC Berkeley." http://www.berkeley.edu/news/media/releases/98legacy/11-23-1998.html.

Merchant, C. 1982. *The Death of Nature: Women, Ecology, and the Scientific Revolution*. New York: Harper & Row.

Merton, R. 1973. "The Normative Structure of Science." In *The Sociology of Science: Theoretical and Empirical Investigations*, ed. N. W. Storer, 267–78. Chicago: University of Chicago Press.

Moffat, A. 1992. "Gene Research Flowers in *Arabidopsis thaliana*." *Science* 258: 1580–81.

Molotch, H. 1976. "The City as a Growth Machine: Toward a Political Economy of Place." *American Journal of Sociology* 82: 309–32.

Monbiot, G. 2002. "The Fake Persuaders: Corporations Are Inventing People to Rubbish Their Opponents on the Internet." *The Guardian* (London), May 14. http://www.guardian.co.uk/internetnews/story/0,7369,715159,00.html

Montgomery, K. C. 1989. *Target: Prime Time*. New York: Oxford University Press.

Morley, D. 1980. *The "Nationwide" Audience*. London: British Film Institute.

Morrill, C. 1991. "Conflict Management, Honor, and Organizational Change." *American Journal of Sociology* 97: 585–621.

Morris, P., and M. Bowden. 2001. "Quinine." In *Robert Burns Woodward: Architect and Artist in the World of Molecules*, ed. O. Benfey and P. Morris, 57–61. Philadelphia: Chemical Heritage Foundation.

Morrow, J., and P. Berg. 1972. "Cleavage of Simian Virus 40 DNA at a Unique Site by a Bacterial Restriction Enzyme." *Proceedings of the National Academy of Sciences* 69: 3365–69.

Mowery, D., R. Nelson, B. Sampat, and A. Ziedonis. 2004. *Ivory Tower and Industrial Innovation: University-Industry Technology Transfer before and after the Bayh-Dole Act in the United States.* Stanford: Stanford University Press.

National Academy of Science. 2003. *Government-Industry Partnerships for the Development of New Technologies.* Washington, D.C.: National Academies Press.

National Academy of Science, Government-University-Industry Research Roundtable. 1988. *Simplified and Standardized Model Agreements for University-Industry Cooperative Research.* Washington, D.C.: National Academies Press.

———. 1993. *Intellectual Property Rights in Industry-Sponsored University Research: A Guide to Alternatives for Research Agreements.* Washington, D.C.: National Academies Press.

National Association of State Universities and Land-Grant Colleges (NASULGC). 1995. *The Land Grant Tradition.* Washington, D.C.: National Association of State Universities and Land-Grant Colleges.

———. 2001. *Shaping the Future: The Economic Impact of Public Universities.* Washington, D.C.: National Association of State Universities and Land-Grant Colleges.

National Institutes of Health. 1994. "Developing Sponsored Research Agreements: Considerations for Recipients of NIH Research and Contracts." In *NIH Guide.* http://grants1.nih.gov/grants/guide/notice-files/not94-213.html.

National Research Council. 1995. *Colleges of Agriculture at the Land Grant Universities: A Profile.* Washington, D.C.: National Academies Press.

———. Committee on the Biological Confinement of Genetically Engineered Organisms. 2004. *Biological Confinement of Genetically Engineered Organisms.* Washington, D.C.: National Academies Press.

———. Committee on Japan. 1992. *U.S.-Japan Technology Linkages in Biotechnology: Challenges for the 1990s.* Washington, D.C.: National Academies Press.

National Science Board. 2004. *Science and Engineering Indicators 2004* (NSB 04-01). Arlington, Va: National Science Foundation, Division of Science Resources Statistics. http://www.nsf.gov/statistics/seind04/.

National Science Foundation. 2000. *Science and Engineering Indicators—2000.* Washington, D.C.: U.S. Government Printing Office.

———. National Science Board. 1978. "Basic Research in the Mission Agencies." Washington, D.C.: National Science Foundation.

———. Office of Legislative and Public Affairs. 2001. "NSF Boosts Funding for Plant Genome Research, November 1, 2001 Press Release No. PR01-89." Arlington, Va.: National Science Foundation, Office of Legislative and Public Affairs.

National Science and Technology Council. Committee on Science, Interagency Working Group on Plant Genomics. 1998. "National Plant Genome Initiative: Final Report." Washington, D.C.: National Science and Technology Council Executive Secretariat.

Nelkin, D., R. R. Nelson, and C. Kiernan. 1987. "Commentary: University-Industry Alliances." *Science, Technology and Human Values* 12: 65–74.

Nelsen, L. 2004. "A US Perspective on Technology Transfer: The Changing Role of the University." *Molecular Cell Biology* 5: 1–5.

Nelson, R. 1993. *National Innovation Systems: A Comparative Analysis.* Oxford: Oxford University Press.

———. 2001. "Observations on the Post–Bayh-Dole Rise of Patenting at American Universities." *Journal of Technology Transfer* 26: 13–19.

Nettelbeck, D. M., and D. T. Curiel. 2003. "Tumor-Busting Viruses." *Scientific American* (September 15): 68–75.

Nordlee, J. A., S. L. Taylor, J. A. Townsend, L. Thomas, and R. K. Bush. 1996. "Identification of a Brazil-Nut Allergen in Transgenic Soybeans." *New England Journal of Medicine* 334: 688–92.

NorthStar Economics, Inc. 2002. "The Economic Impact of the University of Wisconsin, Madison System." Madison: NorthStar Economics, Inc.

Novartis. 2004. "Welcome to Novartis." Basel, Switzerland: Novartis. http://www.novartis.com/.

Novartis Agricultural Discovery Institute, Inc. (NADI), and University of California (UC). 1998. "Collaborative Research Agreement." La Jolla and Oakland: Novartis Agricultural Discovery Institute and Regents of the University of California.

———. 2000a. "Amendment #1 to: Collaborative Research Agreement between Novartis Agricultural Discovery Institute, Inc. and the Regents of the University of California." La Jolla and Oakland: Novartis Agricultural Discovery Institute and Regents of the University of California.

———. 2000b. "Amendment #2 to: Collaborative Research Agreement between Novartis Agricultural Discovery Institute, Inc. and the Regents of the University of California." La Jolla and Oakland: Novartis Agricultural Discovery Institute and Regents of the University of California.

O'Connor, J. 1973. *The Fiscal Crisis of the State*. New York: St. Martin's Press.

Olswang, S., and B. Lee. 1984. *Faculty Freedoms and Institutional Accountability: Interactions and Conflicts*. Washington, D.C.: Association for the Study of Higher Education.

Osborn, D. 1988. *Laboratories of Democracy: A New Breed of Governor Creates Models for National Growth*. Boston: Harvard Business School Press.

Owen, G. 2003. "Dean's Corner—Beating the Odds: Biology and the Budget Deficit." *College News*, September 8. University of California, Berkeley, College of Letters and Science. http://ls.berkeley.edu/new/deanscorner/0309go.html.

Owens, M. J. 1978. "Patent Program of the University of California." In *Patent Policy: Government, Academic, and Industrial Concepts*, ed. W. Marcy, 65–68. Washington, D.C.: American Chemical Society.

Parenti, M. 1993. *Inventing Reality*. New York: St. Martin's Press.

Patrico, J. 2001. "Universities for Sale?" *Progressive Farmer Extra*, November. http://www.progressivefarmer.com/issue/1101/research/default.asp.

Pennock, R. 2003. "Creationism and Intelligent Design." *Annual Review of Genomics and Human Genetics* 4: 143–63.

Pfau, M. R. 1995. "Covering Urban Unrest: The Headline Says It All." *Journal of Urban Affairs* 17 (2): 131–41.

Power, M. 1997. *The Audit Society: Rituals of Verification*. Oxford: Oxford University Press.

Press, E., and J. Washburn. 2000. "The Kept University." *Atlantic Monthly*, March, 39–54.

Price, D. 1979. "The Ethical Principles of Scientific Institutions." *Science, Technology and Human Values* 4: 46–60.

Price, R. M., and L. Goldman. 2002. *The Novartis Agreement: An Appraisal*. Administrative Review, UC Berkeley. October 4. http://www.berkeley.edu/news/media/releases/2004/07/admin_novartis_review.pdf.

Pugh, E. 1984. *Memories That Shaped an Industry: Decisions Leading to IBM System/360*. Cambridge: MIT Press.

Putnam, H. 2002. *The Collapse of the Fact/Value Dichotomy and Other Essays*. Cambridge: Harvard University Press.

———. 2004. *Ethics without Ontology*. Cambridge: Harvard University Press.

Quintella, R. H. 1993. *The Strategic Management of Technology in the Chemical and Petrochemical Industries*. New York: Pinter Publishers.

Quist, D., and I. Chapela. 2001. "Transgenic DNA Introgressed into Traditional Maize Landraces in Oaxaca, Mexico." *Nature* 414 (November 29): 541–43.

Rausser, G. 1998. Memorandum to Chancellor Berdahl on Novartis Proposal, July 19. University of California, Berkeley. In the archives of the UC Berkeley Academic Senate.

Readings, B. 1996. *The University in Ruins*. Cambridge: Harvard University Press.

Reamer, A., L. Icerman, and J. Youtie. 2003. "Technology Transfer and Commercialization: Their Role in Economic Development." Washington, D.C.: U.S. Department of Commerce, Economic Development Administration.

Richardson, L. 1997. "UC Christmas Scrooge." *California Farmer*, December 5.

———. 1998. "National Cancer." *California Farmer*, January 5.

Ricoeur, P. 1977. *The Rule of Metaphor*. Toronto: University of Toronto Press.

Roemer, M. 1970. *Fishing for Growth: Export-led Development in Peru*. Cambridge: Harvard University Press.

———. 1989. "Human Capital and Growth: Theory and Evidence." National Bureau of Economic Research Working Paper 3173. Washington, D.C.: National Bureau of Economic Research, Inc.

Rogers, S. 1970. "Skills for Genetic Engineers." *New Scientist* 45: 194–96.

Rosenberg, C. 1976. *No Other Gods: On Science and American Social Thought*. Baltimore: Johns Hopkins University Press.

Rosenthal, A. 2000. "The Move." *Stanford Medicine* (spring). http://mednews.stanford.edu/stanmed/2000spring/themove.html.

Rosset, P., and M. Moore. 1998. Editorial, "Research Alliance Debated; Deal Benefits Business, Ignores UC's Mission." *San Francisco Chronicle*, October 23, A27.

Rouse, J. 1992. "Independent Research Institutes." In *Innovative Models for University Research*, ed. C. Haden and J. Brink, 115–28. New York: Elsevier Science Publishers.

Roush, W. 1997. "Biology Departments Restructure." *Science* 275: 1556–60.

Sacramento Business Journal. 2002. "Davis Announces Move to Aid Life Sciences." http://www.bizjournals.com/sacramento/stories/2002/12/30/daily38.html.

Sanders, E. 1999. *Roots of Reform: Farmers, Workers, and the American State, 1877–1917*. Chicago: University of Chicago Press.

Sanders, R. 1998. "CNR, Novartis Seal $25 Million Biotech Research Agreement." *The Berkeleyan*, December 2. http://www.berkeley.edu/news/berkeleyan/1998/1202/novartis.html.

———. 2003. "Closing the Book on the Novartis Deal? Internal Campus Study Says Lucrative Agreement Didn't Skew Research or Compromise Academic Freedom." *The Berkeleyan*, January 29. http://www.berkeley.edu/news/berkeleyan/2003/01/29_novart.html.

Sarewitz, D. 1996. *Frontiers of Illusion: Science, Technology and the Politics of Progress*. Philadelphia: Temple University Press.

Schafer, W. 1983. "Normative Finalization." In *Finalization in Science*, ed. W. Schafer, 207–31. Dordrecht: D. Reidel Publishing.

Scheuring, A. F. 1995. *Science and Service: A History of the Land-Grant University and Agriculture in California*. Oakland: University of California, ANR Publications.

Schmidt, C. 2005. "Genetically Modified Foods: Breeding Uncertainty." *Environmental Health Perspectives* 113: A527–33.

Schonbach, P. 1990. *Account Episodes*. New York: Cambridge University Press.

Schurman, R. A., and W. A. Munro. 2003. "Making Biotech History: Social Resistance to Agricultural Biotechnology and the Future of the Biotechnology Industry." In *Engineering Trouble: Biotechnology and Its Discontents*, ed. R. A. Schurman and D. Kelso, 111–29. Berkeley and Los Angeles: University of California Press.

Scott, A., and M. Storper, eds. 1986. *Production, Work, Territory: The Geographical Anatomy of Industrial Capitalism*. Boston: Allen & Unwin.

Securities and Exchange Commission. 1998. "Form 10-K, LECG, Inc.: Annual Report for the Fiscal Year Ending December 31, 1997." http://www.sec.gov/Archives/edgar/data/1047749/0001012870-98-000796.txt.

Selgrade, M., I. Kimber, L. Goldman, and D. Germolec. 2003. "Assessment of Allergenic Potential of Genetically Modified Foods: An Agenda for Future Research." *Environmental Health Perspectives* 111: 1140–41.

Shane, S. 2002. "Selling University Technology: Patterns from MIT." *Management Science* 48: 122–37.

Shanmugam, K., and R. Valentine. 1975. "Molecular Biology of Nitrogen Fixation." *Science* 187: 919–24.

Shapin, S. 2003. "Ivory Trade." *London Review of Books*, September 11, 15–19.

Shapin, S., and S. Schaffer. 1985. *Leviathan and the Air-Pump: Hobbes, Boyle, and the Experimental Life*. Princeton: Princeton University Press.

Sheldon, I. 2004. "Europe's Regulation of Agricultural Biotechnology: Precaution or Trade Distortion?" *Journal of Agricultural and Industrial Organization* 2: 1–26.

Slaughter, S. 2004. *Academic Capitalism and the New Economy*. Baltimore: Johns Hopkins University Press.

Slaughter, S., and L. L. Leslie. 1997. *Academic Capitalism: Politics, Policies, and the Entrepreneurial University*. Baltimore: Johns Hopkins University Press.

Slaughter, S., and G. Rhoades. 1996. "The Emergence of a Competitiveness Research and Development Policy Coalition and the Commercialization of Academic Science and Technology." *Science, Technology and Human Values* 21: 303–39.

Smith, D. E. 1990. *The Conceptual Practices of Power: A Feminist Sociology of Knowledge*. Boston: Northeastern University Press.

Snow, C. P. 1959. *The Two Cultures and the Scientific Revolution*. New York: Cambridge University Press.

Sonnert, G. 2002. *Ivory Bridges: Connecting Science and Society*. Cambridge: MIT Press.

Star, S. L. 1991. "Power, Technology and the Phenomenology of Conventions: On Being Allergic to Onions." In *A Sociology of Monsters: Essays on Power, Technology, and Domination*, ed. J. Law, 26–57. London: Routledge.

Steele, J. 1974. "Biographical Notes: Karl Friedrich Meyer." *Journal of Infectious Diseases* 129 (supplement): S3–9.

Stelljes, K. 2004. "Celebrating Our Past and Our Future: The College of Natural Resources Reaches 30 Years." *Breakthroughs* 10: 3–7.

Stipp, D. 2005. "The Dogged Scientist, the Old Lab Vial, and the Quest to Stop Cancer." *Fortune*, June 27, 175–84.

Storper, M., and R. Salais. 1997. *Worlds of Production: The Action Frameworks of the Economy*. Cambridge: Harvard University Press.

Stratton, K., D. Almario, and M. McCormick, eds. 2002. *Immunization Safety Review: SV40 Contamination of Polio Vaccine and Cancer*. Washington, D.C.: National Academies Press.

Strauss, A. 1978. *Negotiations: Varieties, Contexts, Processes, and Social Order*. San Francisco: Jossey-Bass.

Streicher, S., E. Gurney, and R. Valentine. 1971. "Transduction of the Nitrogen-Fixation Genes in *Klebsiella pneumoniae*." *Proceedings of the National Academy of Sciences* 68: 1174–77.

Strickler, H. 2001. "Simian Virus 40 (SV40) and Human Cancers." *Einstein Quarterly Journal of Biology and Medicine* 18 (1): 14–20.

Students for Responsible Research, University of California, Berkeley. 1998a. "Open Petition to the UC Board of Regents [Novartis Agreement]." In the archives of the UC Berkeley Academic Senate.

———. 1998b. "Media Release: 'Angry Students Denounce Proposed $50 Million Alliance between Biotech Giant Novartis and UC Berkeley's College of Natural Resources.'" November 23. http://list.jca.apc.org/public/asia-apec/1998-November/000908.html.

———. N.d. "The College of Natural Resources–Novartis Alliance." http://nature.berkeley.edu/srr/index.htm.

Syngenta. 2002. "Syngenta and Diversa Form Extensive Research and Product Development Alliance." News Release. Basel, Switzerland, and San Diego: Syngenta. http://www.syngenta.com/en/media/article.aspx?pr=120402&Lang=en.

———. 2003–2005. Annual Reports. http://www.syngenta.com.

―――. 2004. "Form 20-F, Syngenta AG, Annual filing with the Securities and Exchange Commission for the Fiscal Year Ending December 31, 2003." http://www.sec.gov/Archives/edgar/data/1123661/000095010304000439/mar1804_20f.htm.

Tait, J., and J. Chataway. 2000. "Novartis Agribusiness Monograph." In *PITA Project: Policy Influences on Technology for Agriculture: Chemicals, Biotechnology and Seeds*, ed. J. Tait, J. Chataway and David Weild. Annex 22. Edinburgh: Scottish Universities Policy Research and Advice Network.

Ten Eyck, T. A. 2000. "Interpersonal and Mass Communication: Matters of Trust and Control." *Current Research in Social Psychology* 5 (14): 206–24.

Ten Eyck, T. A. and F. Deseran. 2004. "Oyster Coverage: Chiastic News as a Reflection of Local Expertise and Economic Concerns." *Sociological Research Online* 9. http://www.socresonline.org.uk/9/4/ten_eyck.html.

Thayer, A. 2002. "DuPont, Monsanto Bury the Hatchet." *Chemical and Engineering News* 80: 9.

Thévenot, L. 2001. "Organized Complexity: Conventions of Coordination and the Composition of Economic Arrangements." *European Journal of Social Theory* 4 (4): 405–25.

Thompson, D. 2000. "Universities Criticized for Research Contracts with Private Firms." Associated Press, May 16. http://www.biotech-info.net/universities_criticized.html.

Thompson, J. 1990. *Ideology and Modern Culture*. Stanford: Stanford University Press.

Thurow, L. C. 1999. *Building Wealth: The New Rules for Individuals, Companies, and Nations in a Knowledge-Based Economy*. New York: HarperCollins.

Thursby, J. G., R. Jensen, and M. C. Thursby. 2001. "Objectives, Characteristics and Outcomes of University Licensing: A Survey of Major U.S. Universities." *Journal of Technology Transfer* 26 (1–2): 59–72.

Torrey Mesa Research Institute (TMRI) and University of California (UC). 2002. "Amendment #3 to: Collaborative Research Agreement No. 010134." La Jolla and Oakland: Torrey Mesa Research Institute and Regents of the University of California.

Trow, M. 2004. "Leadership and Academic Reform: Biology at Berkeley." http://www.ishi.lib.berkeley.edu/cshe/mtrow.

Tuchman, G. 1978. *Making News*. New York: Free Press.

Turbayne, C. M. 1970. *The Myth of Metaphor*. Columbia: University of South Carolina Press.

University of California. 1998. "Questions and Answers on the UC Berkeley–NADI Research Agreement." Press Release, November 23. http://www.berkeley.edu/news/media/releases/98legacy/QandA.html.

―――. 2002. "Five Years of Progress: A Summary Report on the Results of the 1997 President's Retreat; the University of California's Relationships with Industry in Research and Technology Transfer." Oakland: University of California.

―――. 2003. "General University Policy Regarding Academic Appointees." Academic Personnel Manual. http://www.ucop.edu/acadadv/acadpers/apm/section1.pdf.

―――. 2004. Digital Archives. http://sunsite.berkeley.edu/uchistory/.

―――. Office of the President. 1989. "Guidelines on University-Industry Relations." Oakland: University of California.

―――. 1990. "University Policy on Integrity in Research." Oakland: University of California. http://www.ucop.edu/ucophome/coordrev/policy/6-19-90.html.

―――. 2003. "Guidance for Faculty and Other Academic Employees on Issues Related to Intellectual Property and Consulting." Oakland: University of California.

University of California, Berkeley. 2002. "A Brief History of the University." http://www.berkeley.edu/about/history.

―――. 2003a. "Berkeley Division of the Academic Senate's Comments on the Strategic Academic Plan." http://www.berkeley.edu/news/media/releases/2003/05/sap/comments.shtml (accessed July 1, 2005).

————. 2003b. "The Pulse of Scientific Freedom in the Age of the Biotech Industry." http://webcast.berkeley.edu/events.

————. 2005. "Principles of Community." http://www.berkeley.edu/about/community.shtml (accessed July 1, 2005).

————. Academic Senate. 1998. "Issue of Outside Work for Faculty Revived in Concern over 'Conflict of Commitment.'" *Notice: A Publication of the Academic Senate* 22 (1): 3.

————. College of Natural Resources (CNR). 2004. College of Natural Resources at the University of California, Berkeley. http://www.cnr.berkeley.edu/site/index.php.

————. College of Natural Resources Executive Committee (CNR ExCom). 1999. "Results of the Second (Modified) Faculty Survey on the Strategic Alliance of the College of Natural Resources (CNR) with Novartis." University of California, Berkeley.

————. Department of Plant and Microbial Biology (PMB). 2002. Website of the Department of Plant and Microbial Biology, University of California, Berkeley. http://mollie.berkeley.edu/.

————. Department of Plant and Microbial Biology (PMB) Graduate Students. 1998. Letter to PMB faculty, December 14. University of California, Berkeley. In the archives of the UC Berkeley Academic Senate.

————. Divisional Council of the Academic Senate (DIVCO). 2003. "Enclosures #2B-2E for the Approved Minutes of the Berkeley Divisional Council." February 10. University of California, Berkeley. In the archives of the UC Berkeley Academic Senate.

————. Office of Technology Transfer. 2001. "Guidance for Faculty and Other Academic Employees on Issues Related to Intellectual Property and Consulting." Berkeley: University of California.

————. Public Information Office. 1998. "Questions and Answers on the UC-Berkeley–NADI Research Agreement." Press Release, November 23. http://www.berkeley.edu/news/media/releases/98legacy/QandA.html.

————. Sponsored Projects Office. 1998–2003. "University of California, Berkeley: Sponsored Projects Annual Report, Fiscal Year 1998." http://coeus.spo.berkeley.edu/guest_report.asp.

————. Strategic Planning Committee. 2002. "U.C. Berkeley Strategic Academic Plan." http://www.berkeley.edu/news/media/releases/2003/05/sap/plan.pdf.

Upshaw, C. 2000. Novartis Discussion Summary (e-mail). University of California, Berkeley. In the archives of the UC Berkeley Academic Senate.

U.S. Congress. Office of Technology Assessment. 1985. "Technology, Public Policy, and the Changing Structure of American Agriculture: A Special Report for the 1985 Farm Bill." Washington, D.C.: U.S. Government Printing Office.

————. 1991. "Biotechnology in a Global Economy." Washington, D.C.: U.S. Government Printing Office.

U.S. Congress. Senate. Committee on Commerce, Science, and Transportation. 1978. "Hearings before the Subcommittee on Science, Technology, and Space of the Committee on Commerce, Science, and Transportation, United States Senate, Ninety-Fifth Congress: First Session on Regulation of Recombinant DNA Research, November 2, 8 and 10, 1977." Washington, D.C.: U.S. Government Printing Office.

U.S. Department of Commerce. International Trade Administration. 1984. "Biotechnology: High Technology Industries: Profiles and Outlooks." Washington, D.C.: U.S. Department of Commerce, International Trade Administration.

U.S. Department of Health and Human Services. 2004. *Final Guidance Document: Financial Relationships and Interests in Research Involving Human Subjects; Guidance for Human Subject Protection.* May 5. http://www.hhs.gov/ohrp/humansubjects/finreltn/fguid.pdf.

U.S. General Accounting Office. 2001. "Biomedical Research: HHS Direction Needed to Address Financial Conflicts of Interest." GAO-02-89. Washington, D.C.: U.S. General Accounting Office.

————. 2002. "Genetically Modified Foods: Experts of Safety Tests as Adequate, But FDA's Evaluation Process Could Be Enhanced." Washington, D.C.: U.S. General Accounting Office.

Varga, A. 1998. *University Research and Regional Innovation: A Spatial Econometric Analysis of Academic Technology Transfers*. Boston: Kluwer Academic Publishers.

Vest, C. M. 1996. "'Not What We Think, What We Haven't Thought Of.'" Keynote address at the 1996 Jerome B. Wiesner Symposium: The Future of the Government/University Partnership, University of Michigan, Ann Arbor. http://web.mit.edu/president/communications/JBWSymp-2-96.html.

————. 2005. *Pursuing the Endless Frontier*. Cambridge: MIT Press.

Vettel, E. 2004. "The Protean Nature of Stanford University's Biological Sciences, 1946–1972." *Historical Studies in the Physical and Biological Sciences* 35 (1): 95–113.

Vogel, G. 2002. "Retreat from Torrey Mesa: A Chill Wind in Ag Research." *Science* 298 (5601): 2106.

Wade, N. 1972. "Division of Biologics Standards: The Boat That Never Rocked." *Science* 175 (4027): 1225–30.

Walsh, S. 2004. "Berkeley Denies Tenure to Ecologist Who Criticized University's Ties to the Biotechnology Industry." *Chronicle of Higher Education*, January 9, A10.

Walzer, M. 1983. *Spheres of Justice: A Defense of Pluralism and Equality*. New York: Basic Books.

Washburn, J. 2005. *University, Inc.: The Corporate Corruption of Higher Education*. New York: Basic Books.

Weaver, W., C. P. Snow, T. Hesburg, and W. Baker. 1961. "The Moral Un-Neutrality of Science." *Science* 133 (3448): 255–62.

Weber, M. 1947. *The Theory of Social and Economic Organization*. New York: Oxford University Press.

Webster, A. and H. Etzkowitz. 1998. "Toward a Theoretical Analysis of Academia-Industry Collaboration." In *Capitalizing Knowledge: New Intersections of Industry and Academia*, ed. A. Webster, P. Healey, and H. Etzkowitz, 47–72. Albany: State University of New York Press.

Wheelwright, P. 1954. *The Burning Fountain*. Bloomington: Indiana University Press.

Whitehead, D. 1968. *The Dow Story: The History of the Dow Chemical Company*. New York: McGraw-Hill.

Williams-Jones, B. 2005. "Knowledge Commons or Economic Engine—What's a University For?" *Journal of Medical Ethics* 31 (5): 249–50.

Yamada, K., et al. 2003. "Empirical Analysis of Transcriptional Activity in the *Arabidopsis* Genome." *Science* 302 (5646): 842–46.

Yano, H., J. Wong, Y. Lee, M. Cho, and B. Buchanan. 2001. "A Strategy for the Identification of Proteins Targeted by Thioredoxin." *Proceedings of the National Academy of Sciences* 98: 4794–99.

Yost, J. 2003. "How Ethics Can Advance High-Tech Linkages between Universities and Industries." *Research Management Review* 13 (1): 31–43.

Zelizer, B. 1992. *Covering the Body*. Chicago: University of Chicago Press.

————. 1993. "Journalists as Interpretive Communities." *Critical Studies in Mass Communication* 10 (1): 219–37.

Zucker, L., M. Darby, and J. Armstrong. 2002. "Commercializing Knowledge: University Science, Knowledge Capture, and Firm Performance in Biotechnology." *Management Science* 48 (1): 138–53.

INDEX

academic capitalism, 24, 25–27, 28, 31, 164
Academic Senate Committee on Research, UC Berkeley (COR), 57, 60, 66
Academic Senate Divisional Council, UC Berkeley (DIVCO), 57, 58, 65, 66, 109
accountability, 197–205
actor-network theory, 1
Adams, Roger, 190
Adams Act, 38
adjunct faculty, 53, 54, 77, 78, 82, 134
AgBioWorld, 68
agribusiness, 36, 50, 75, 81, 160, 161
Altieri, Miguel, 75
American Association of University Professors, 6, 7, 99, 150, 206
Amundson, Ronald, 59
antifreeze tomato, 79
Arabidopsis, 13, 83, 127, 187, 214
asymmetrical convergence, 24, 29, 31
Atkinson, Richard, 75, 173
Atlantic Monthly 103
atrazine, 70
"auction" process, 48–49, 54
audit society, 29–30
autonomy, 4, 6–9, 16, 24, 31, 38, 54, 60, 74, 75, 76, 123, 179, 180, 184, 193, 197, 199, 204

Bacillus thuringiensis (Bt), 50, 201
Bacon, Francis, 10, 11

Bay Guardian, The San Francisco, 100
Bayh-Dole Act, 22, 86, 90, 141, 147, 148, 183, 203
Beissinger, Steven, 68
Berdahl, Robert, 58, 61, 98, 107
Berg, Paul, 184, 185, 186
Berkeleyan, The, 97–98
Bethke, Paul, 117, 119
BioSTAR, 145
Biotechnology Planning Board, 46, 62
Birgeneau, Robert, 69, 107
Bok, Derek, 24, 45, 164, 165, 166
Boltanski, Luc, 1–4
Bourdieu, Pierre, 8, 9, 13
Bowker, Albert H., 43
Boyer, Herbert W., 85, 141
Boyle, Robert, 10
Brentano, Robert, 57, 58, 62
Briggs, Steven, 50, 51, 52, 58, 65, 112, 113, 189
Bright, Simon, 112, 114, 147
Buchanan, Bob, xiii, 47, 54, 118, 120, 144, 145, 202
Burbank, Luther, 138
Bush, Vannevar, 19, 20, 24

C4 (Committee of Four), xiii, 47, 48, 57, 62, 109
Calgene, 47, 79, 85

capitalism, academic, 24, 25–27, 28, 31, 164
Center for Studies in Higher Education, UC Berkeley (CSHE), 58, 60, 61, 62, 63, 116
cephalosporin, 182
Cerny, Joseph, 46, 52, 58, 63
Chapela, Ignacio, x, 56, 67–69, 107
Christ, Carol, 57, 58, 60, 61, 62, 63, 65
chronology, 46–70
Ciba-Geigy, 49, 50, 74, 112, 182, 189
College of Natural Resources, UC Berkeley, Executive Committee (ExCom), 56, 57, 59, 62, 67, 77
collegiality, 135, 153–55, 162
Committee of Four (C4), xiii, 47, 48, 57, 62, 109
conflict of interest, 69, 70, 173–78, 201, 202, 213, 214
conventions theory, 1–12, 16, 193–95
convergence, asymmetrical, 24, 29, 31
Cooperative Research and Development Agreement, USDA (CRADA), 54
COR (UC Berkeley Academic Senate Committee on Research), 57, 60, 66
Cottrell, Frederick Gardner, 140–41
Crabb, Mary, 61, 116
CRADA (USDA Cooperative Research and Development Agreement), 54
creativity, 4–6, 9, 24, 31, 54, 60, 78, 85, 96, 100, 180
creativity, autonomy, and diversity, 4–12, 13, 16, 17, 22, 30, 36, 93, 95, 162, 166, 179, 197, 198, 206
CSHE (UC Berkeley Center for Studies in Higher Education), 58, 60, 61, 62, 63, 116
culture wars, 32, 157, 163, 165

Daiichi Pharmaceutical Corporation, 53
Daily Californian, The, 100
Day, Boysie, 39
Department of Energy, US, 20, 23, 76, 125
desserts, 13, 14, 17
Diamond v. Chakrabarty, 22, 139, 203
DIVCO (UC Berkeley Divisional Council of the Academic Senate), 57, 58, 65, 66, 109
Diversa, 64–65, 115, 189
diversity, 9–12, 24, 34, 54, 60, 78, 81, 82, 135, 152, 153, 157, 162, 166, 173, 180, 197, 199, 200
Dow Chemical Company, 50, 65, 190

DuPont, 49, 50, 65, 110, 112, 140, 199, 200
DuPont/Pioneer, 52

engines of growth, universities as, 166–170, 205
entrepreneurial, 22, 25, 27, 164, 173, 189
ethics, 181, 183–84, 187–88, 192, 194, 195–97, 203–5. *See also* integrity
ExCom (Executive Committee of the UC Berkeley College of Natural Resources), 56, 57, 59, 62, 67, 77

Felde, Marie, 73
frankenfoods, 10
Freedman, Joyce, 52

Genentech, 85
Getz, Wayne, 69
Gill Tract, 152
Golan, Talila, 113
Gomes, W. R., 160
governance, shared, ix, 73, 154, 166
Gray, Paul, 61, 63, 77, 160
Gruissem, Wilhelm, xiii, 46–47, 62, 63, 66, 110, 111, 115, 116, 123, 124
Gutierrez, Andrew, 56

Hackett, E. J., 28–29, 31
Handelsman, Jo, 175
Haraway, Donna, 10
Hardin, Charles, 8
Harding, Sandra, 7, 11, 16
Hatch Act, 37, 41
Hayden, Tom, 59
Hayes, Tyrone, 67, 69–70
Henson, Dave, 78, 161
Heyman, Michael, 61
Hilgard, E. W., 41
Hitachi Chemical Company, 53, 85
Hoechst, 85
Hoskins, William, 46, 52–53, 58, 62, 118
Humboldtian perspectives on higher education, 11, 19, 32, 33, 164–65, 198, 206

integrity, individual, 195–96
integrity and accountability, 197–205
intellectual property, 22, 23, 28, 39, 47, 76, 87, 137–77, 189, 190, 192, 193, 195, 202, 205
International Biotechnology Advisory Board, 47, 62

Jones, Russell, 110, 111, 119
justifications, 2, 3, 12, 34, 45, 104, 105,
    110–11

Kantian perspectives on higher education,
    1, 11, 32, 33, 164–65, 198, 206
Kerr, Clark, 157, 193, 213
Kleinman, Daniel, 29, 31, 175–76, 214
Kornberg, Arthur, 85, 186
Krimsky, Sheldon, 25, 27, 31
Kroken, Scott, 113, 115, 124
Krueger, Paul, 185
Kustu, Sydney, 112, 114, 118, 121, 125, 135

land grant mission, 38, 40–43, 152, 159–61,
    186
LaPorte, Todd, 57
Latour, Bruno, 67
Lave, Jean, 60, 61, 62
LECG (Law and Economics Consulting
    Group), 74–75
Leister, Todd, 117
Lemaux, Peggy, xiii, 47
Ludden, Paul, 147

MacLachlan, Anne, 60, 61, 63, 116
Madey, John M. J. v. Duke University, 26,
    143–44
Malkin, Richard, 61, 63, 69, 120, 146
Marrs, Barry L., 189
McQuade, Donald, 155
Meyer, Karl Friedrick, 185
Mimura, Carol, 47, 53, 58, 62
Monsanto, 47, 49, 50, 51, 64, 68, 74, 84, 85,
    110, 112, 140, 182, 189
Morrill Act, 36, 40, 96, 156, 171
multiversity, 4, 18, 32, 33, 36, 157, 166, 193

Natural Resources, College of, UC Berkeley,
    Executive Committee (ExCom), 56, 57,
    59, 62, 67, 77
Nature, x, 67, 68
Noble, David, 99

OTL (UC Berkeley Office of Technology
    Licensing), 46, 47, 52, 53, 58, 66, 76, 77,
    118, 143, 146, 181, 194, 195, 211
Ottinger, Gwen, 60
Oversight Committee for the UC Berkeley–
    Novartis Agreement, 61, 63, 155

Pandora's box, 101, 154, 203
PGEC (USDA Plant Gene Expression Cen-
    ter), 53–54, 146
Pioneer/Pioneer Hi-Bred, 49, 52, 110, 112,
    113, 140, 144, 189
Plant Biology, Department of, 43, 44
Plant Gene Expression Center, USDA
    (PGEC), 53–54, 146
Plant Patent Act of 1930 (PPA), 138–39
Plant Variety Protection Act (PVPA), 138,
    139, 140
populism/populist, 35, 36, 37, 38, 39, 42, 45,
    79, 81, 84, 157, 158, 159, 161, 167, 169,
    194, 198
postdoctoral scholars, 73, 109, 113, 115,
    116, 117, 123, 124, 126, 128, 182
Power, Michael, 29–31, 34, 174
PPA (Plant Patent Act of 1930), 138–39
Price Report, 127, 170
progressivism/progressives, 35–36, 37, 38,
    39–40, 42, 43, 45, 79, 84, 101, 156, 163,
    167, 194, 198
PVPA (Plant Variety Protection Act), 138,
    139, 140

Quail, Peter, 54, 113
Quist, David, 55, 67, 68, 152

Rausser, Gordon, xiii, 47–49, 55, 56, 57,
    58, 61, 62, 63, 72, 73, 74–75, 92, 98, 99,
    109, 152
Readings, Bill, 32–33, 164, 165, 166
Rine, Jasper, 69
Rosset, Peter, 75, 81, 155, 158, 161

San Francisco Bay Guardian, The, 100
Sanders, Robert, 73
Schultz, T. W., 8
science wars, x, 157, 163, 165
Scripps-Sandoz Agreement, 46, 89
shared governance, ix, 73, 154, 166
Slaughter, Sheila, 26–28, 31
Smith-Lever Act, 38, 88
Spear, Robert, 57, 61, 63
Spheres of Justice (Michael Walzer), 13
SPO (UC Berkeley Sponsored Projects
    Office), 52, 63, 73, 194
SRR (Students for Responsible Research),
    54, 55–56, 58–59, 71, 74–75, 78, 98
Staley, Cady, 190

Staskawicz, Brian, 63, 66
Sung, Renee, 126
Syngenta, x, 63, 64–65, 67, 70, 99, 112, 114,
  145, 147, 202, 210, 213

Taylor, Paul, 42
Teitz, Michael, 61, 63
tests and trials, 12–15
Theologis, Athanasios, 54
Thévenot, Laurent, 1–4
Tien, Chang-Lin (UC Berkeley Chancellor),
  46, 62
tomato, antifreeze, 79

UCOP (University of California Office of the
  President), 47, 58, 196, 197
USDA Cooperative Research and Develop-
  ment Agreement (CRADA), 54
USDA Plant Gene Expression Center
  (PGEC), 53–54, 146

Vorih, Jennifer, 117

Walzer, Michael, 11, 13, 14
Woodward Research Institute, 181–83

Zambryski, Patricia, 112

ALAN RUDY teaches in the Department of Sociology, Anthropology and Social Work at Central Michigan University.

DAWN COPPIN is the Executive Director of the Homeless Garden Project in Santa Cruz, California.

JASON KONEFAL is a Doctoral Candidate, Department of Sociology, Michigan State University.

BRADLEY T. SHAW is Licensing and Marketing Manager in the Office of Intellectual Property at Michigan State University.

TOBY A. TEN EYCK is Associate Professor, Department of Sociology and the National Food Safety and Toxicology Center, Michigan State University.

CRAIG K. HARRIS is Associate Professor, the Department of Sociology and the Food Safety Policy Center, Michigan State University.

LAWRENCE BUSCH is Director, Institute for Food and Agricultural Standards and University Distinguished Professor, Michigan State University.